Preface

Welcome to our journey into the fascinating world of trigonometry, a branch of mathematics that has gratified the minds of scholars, scientists, and mathematicians for centuries. This textbook is designed to introduce you to the fundamental concepts, principles, and applications of trigonometry. It contains a comprehensive guide that will support your learning and understanding of this essential subject.

Trigonometry comes from the Greek words "trigonon" (triangle) and "metron" (measure). It is the study of the relationships between the angles and sides of triangles, particularly right-angled triangles. It is a subject that extends beyond the pure mathematics. The ability to understand and manipulate trigonometric functions is a powerful tool, opening doors to solving real-world problems and understanding the universe around us.

This textbook is structured to cater to learners at different levels, starting with the basics of trigonometry and gradually progressing to more complex concepts. Each chapter is carefully crafted to build upon the previous one, ensuring a smooth and logical progression through the material. We begin with an introduction to the definition of trigonometric functions in a right-angled triangle, followed by explorations of the unit circle, trigonometric identities and equations.

Our approach is to blend theoretical knowledge with practical application. Each chapter includes a variety of examples, and exercises that not only reinforce the concepts learned but also demonstrate the relevance of trigonometry in solving practical problems.

Whether you are a student encountering trigonometry for the first time, a teacher looking for a comprehensive resource, or someone with a general interest in mathematics, this textbook is intended to meet your needs. Our goal is to make trigonometry accessible, understandable, and, most importantly, enjoyable.

As you embark on this mathematical journey, we invite you to approach the subject with curiosity and an open mind. The study of trigonometry is not just about learning mathematical formulas; it is about developing a new way of thinking and seeing the world. We hope that this textbook will be a valuable companion on your journey, providing you with the knowledge and tools to explore the fascinating world of trigonometry.

Table of Contents

1   Think in a Right Triangle   3

2   The values of Trigonometric Functions of Special Angles   13
    2.1   Trigonometric functions of angle $45°$ 13
    2.2   Trigonometric functions of angles $30°$ and $60°$ 14
    2.3   Trigonometric Functions of Special Angles 16

3   Measure of Angles   19
    3.1   Radian 19
    3.2   Conversion between radians and degrees 20

4   Trigonometric Functions in the Unit Circle   23
    4.1   Symmetry, shifts, and periodicity 26
        4.1.1   Symmetry 26
        4.1.2   Shifts and periodicity 28

5   Trigonometric Identities   31
    5.1   Addition and Difference Formulas 31
        5.1.1 $\cos(\alpha - \beta)$ 31
        5.1.2 $\cos(\alpha + \beta)$ 32
        5.1.3 $\sin(\alpha - \beta)$ 33
        5.1.4 $\sin(\alpha + \beta)$ 34
        5.1.5 $\tan(\alpha - \beta)$ 35
        5.1.6 $\tan(\alpha + \beta)$ 36
    5.2   Double-Angle Formulas 38
    5.3   Half-Angle Formulas 42
    5.4 $\sin x$, $\cos x$ and $\tan x$ in terms of $t = \tan \frac{x}{2}$ 44
    5.5   Product to Sum Formulas 46

    5.6   Sum to Product Formulas 48

6 Trigonometric Equations   51
    6.1   Sines 51
    6.2   Cosines 53
    6.3   tangent 55

7 Problems   57

8 Solutions   71

Chapter 1

Think in a Right Triangle

We start the first chapter of this book by introducing the definition of trigonometric functions. Basically, to be familiar with trigonometric functions, we have to know the definition of trigonometric functions. The starting point of trigonometric functions was mainly consider in a right triangle. Before going to the definitions of trigonometric functions, the readers should know about the terminology of right triangles.

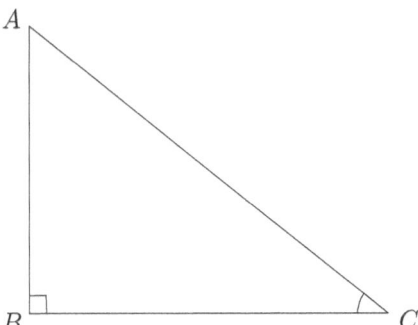

With respect to angle C, AB and BC are called the opposite side and adjacent side respectively. Generally, the opposite side of the right angle is called the hypotenuse.

Chapter 1. Think in a Right Triangle

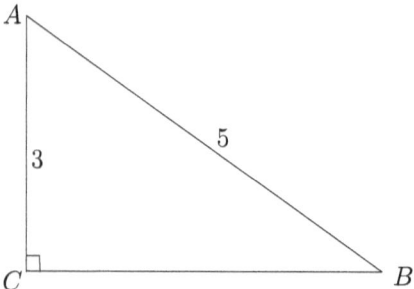

Figure 1.1: Figure of Example 1

Let ABC be a triangle right angled at C. We define the definitions of trigonometric functions below:

$$\sin A = \frac{BC}{AC} = \frac{\text{Opposite side of angle A}}{\text{Hypotenuse}},$$
$$\cos A = \frac{AB}{AC} = \frac{\text{Adjacent side of angle A}}{\text{Hypotenuse}},$$
$$\tan A = \frac{BC}{AB} = \frac{\text{Opposite side of angle A}}{\text{Adjacent side of angle A}},$$

and
$$\cot A = \frac{BC}{AC} = \frac{\text{Adjacent side of angle A}}{\text{Opposite side of angle A}}.$$

Note that sin, cos, tan and cot stand for sine, cosine, tangent and cotangent respectively. They are called trigonometric functions.

> **Example 1**
> Let ABC be a triangle right angled at C. Given that $AB = 5$, $AC = 3$. Determine $\sin A$, $\cos A$, $\tan A$ and $\cot A$.

Solution. Since ABC is a triangle right angled at C, using Pythagorean theorem, we obtain $AB^2 = AC^2 + BC^2$. From the hypothesis, $AB = 5$ and $AC = 3$. Then $5^2 = 3^2 + BC^2$. It implies that $BC^2 = 5^2 - 3^2 = 25 - 9 = 16$. It follows that $BC = \sqrt{16} = \sqrt{4^2} = 4$.

Consequently,

$$\sin A = \frac{BC}{AB} = \frac{4}{5},$$
$$\cos A = \frac{AC}{AB} = \frac{3}{5},$$
$$\tan A = \frac{BC}{AC} = \frac{4}{3},$$
and $\quad \cot A = \frac{AC}{BC} = \frac{3}{4}.$

> **Example 2**
> Let ABC be a triangle right angled at C. Given that $AC = 6$ and $BC = 8$. Determine $\sin A$, $\cos A$, $\tan A$ and $\cot A$.

Solution. Since ABC is a triangle right angled at C, we obtain

$$\begin{aligned} AB^2 &= AC^2 + BC^2 \quad \text{(Pythagorean theorem)} \\ &= 6^2 + 8^2 \\ &= 36 + 64 \\ &= 100 \\ &= 10^2. \end{aligned}$$

It follows that $AB = 10$. It turns out that

$$\sin A = \frac{BC}{AB} = \frac{8}{10} = \frac{4}{5},$$
$$\cos A = \frac{AC}{AB} = \frac{6}{10} = \frac{3}{5},$$
$$\tan A = \frac{BC}{AC} = \frac{8}{6} = \frac{4}{3},$$
and $\quad \cot A = \frac{AC}{BC} = \frac{6}{8} = \frac{3}{4}.$

Observe that the values of $\sin A, \cos A, \tan A$ and $\cot A$ do not depend on the lenghts of the sides of triangles. It depends only on the angles. Hence, the above definition is well-defined.

Chapter 1. Think in a Right Triangle

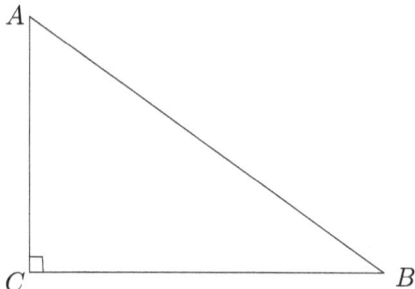

Figure 1.2: Figure of Example 1

Example 3

In a triangle ABC right angled at C, Prove that $\tan A = \dfrac{\sin A}{\cos A}$ and $\cot A = \dfrac{\cos A}{\sin A}$.

Solution. From the definitions of trigonometric functions, We have $\tan A = \dfrac{BC}{AC} = \dfrac{\frac{BC}{AB}}{\frac{AC}{AB}} = \dfrac{\sin A}{\cos A}$.

Additionally, $\cot A = \dfrac{AC}{BC} = \dfrac{\frac{AC}{AB}}{\frac{BC}{AB}} = \dfrac{\cos A}{\sin A}$.

Example 4

Let ABC be a triangle right angled at C. Prove that

1. $\sin^2 A + \cos^2 A = 1$;

2. $1 + \tan^2 A = \dfrac{1}{\cos^2 A}$;

3. $1 + \cot^2 A = \dfrac{1}{\sin^2 A}$;

4. $\cot A = \dfrac{1}{\tan A}$.

Solution. Prove that

1. $\sin^2 A + \cos^2 A = 1$

    Using Pythagorean theorem, $AB^2 = AC^2 + BC^2$. Divide both side of the equality by AB^2, we obtain
    $$1 = \frac{AC^2}{AB^2} + \frac{BC^2}{AB^2}$$
    $$= \left(\frac{AC}{AB}\right)^2 + \left(\frac{BC}{AB}\right)^2$$
    $$= \cos^2 A + \sin^2 A.$$
    Consequently, $\sin^2 A + \cos^2 A = 1$.

2. $1 + \tan^2 A = \dfrac{1}{\cos^2 A}$

    From (1), $\sin^2 A + \cos^2 A = 1$. Then
    $$\frac{\sin^2 A}{\cos^2 A} + \frac{\cos^2 A}{\cos^2 A} = \frac{1}{\cos^2 A}$$
    or
    $$\left(\frac{\sin A}{\cos A}\right)^2 + 1 = \frac{1}{\cos^2 A}.$$
    Therefore, $1 + \tan^2 A = \dfrac{1}{\cos^2 A}$.

3. $1 + \cot^2 A = \dfrac{1}{\sin^2 A}$

    We have $\sin^2 A + \cos^2 A = 1$. It follows that
    $$\frac{\sin^2 A}{\sin^2 A} + \frac{\cos^2 A}{\sin^2 A} = \frac{1}{\sin^2 A}$$
    or
    $$1 + \left(\frac{\cos A}{\sin A}\right)^2 = \frac{1}{\sin^2 A}.$$
    Thus, $1 + \cot^2 A = \dfrac{1}{\sin^2 A}$.

4. $\cot A = \dfrac{1}{\tan A}$

    We have $\cot A = \dfrac{\cos A}{\sin A} = \dfrac{1}{\dfrac{\sin A}{\cos A}} = \dfrac{1}{\tan A}.$

    Therefore, $\cot A = \dfrac{1}{\tan A}$.

> **Example 5**
>
> Given ABC is a triangle right angled at C such that $\sin A = \dfrac{1}{3}$. Find $\cos A, \tan A$ and $\cot A$.

Solution. Find $\cos A, \tan A$ and $\cot A$.
We know that $\sin^2 A + \cos^2 A = 1$.
It follows that $\cos^2 A = 1 - \sin^2 A$.
Since $\sin A = \dfrac{1}{3}$, we obtain

$$\cos^2 A = 1 - \left(\dfrac{1}{3}\right)^2$$
$$= 1 - \dfrac{1}{9}$$
$$= \dfrac{8}{9}.$$

It implies that $\cos A = \dfrac{2\sqrt{2}}{3}$.
Consequently,

- $\tan A = \dfrac{\sin A}{\cos A} = \dfrac{\frac{1}{3}}{\frac{2\sqrt{2}}{3}} = \dfrac{1}{2\sqrt{2}} = \dfrac{\sqrt{2}}{4}$

- $\cot A = \dfrac{1}{\tan A} = \dfrac{1}{\frac{\sqrt{2}}{4}} = \dfrac{4}{\sqrt{2}} = 2\sqrt{2}$

Therefore, $\cos A = \dfrac{2\sqrt{2}}{3}, \tan A = \dfrac{\sqrt{2}}{4}$ and $\cot A = 2\sqrt{2}$.

> **Example 6**
>
> Given ABC is a triangle right angled at C such that $\cos A = \dfrac{2}{3}$. Find $\sin A, \tan A$ and $\cot A$.

Solution. Find $\sin A, \tan A$ and $\cot A$.
We have $\sin^2 A + \cos^2 A = 1$.

Then $\sin^2 A = 1 - \cos^2 A$.
By knowing that $\cos A = \dfrac{2}{3}$, we obtain

$$\sin^2 A = 1 - \left(\dfrac{2}{3}\right)^2$$
$$= 1 - \dfrac{4}{9}$$
$$= \dfrac{5}{9}.$$

It follows that $\sin A = \dfrac{\sqrt{5}}{3}$.
Hence,

- $\tan A = \dfrac{\sin A}{\cos A} = \dfrac{\frac{\sqrt{5}}{3}}{\frac{2}{3}} = \dfrac{\sqrt{5}}{3} \times \dfrac{3}{2} = \dfrac{\sqrt{5}}{2}$

- $\cot A = \dfrac{1}{\tan A} = \dfrac{1}{\frac{\sqrt{5}}{2}} = \dfrac{2}{\sqrt{5}} = \dfrac{2\sqrt{5}}{5}$

> **Example 7**
> Given ABC is a triangle right angled at C such that $\tan A = \dfrac{1}{4}$. Find $\sin A, \cos A$ and $\cot A$.

Solution. Find $\sin A, \cos A$ and $\cot A$.
We know that $1 + \tan^2 A = \dfrac{1}{\cos^2 A}$.
It follows that $\cos^2 A = \dfrac{1}{1 + \tan^2 A}$.
By knowing that $\tan A = \dfrac{1}{4}$, we obtain

$$\cos^2 A = \dfrac{1}{1 + \left(\dfrac{1}{4}\right)^2}$$
$$= \dfrac{1}{1 + \dfrac{1}{16}}$$

9

$$= \frac{1}{\frac{17}{16}}$$
$$= \frac{16}{17}.$$

It implies that $\cos A = \dfrac{4}{\sqrt{17}} = \dfrac{4\sqrt{17}}{17}$.
Moreover, $\sin^2 A + \cos^2 A = 1$.
Then
$$\sin^2 A = 1 - \cos^2 A$$
$$= 1 - \left(\frac{4}{\sqrt{17}}\right)^2$$
$$= 1 - \frac{16}{17}$$
$$= \frac{1}{17}.$$

Hence, $\sin A = \dfrac{1}{\sqrt{17}} = \dfrac{\sqrt{17}}{17}$.
It follows that $\cot A = \dfrac{1}{\tan A} = \dfrac{1}{\frac{1}{4}} = 4$.
Therefore, $\cos A = \dfrac{4\sqrt{17}}{17}, \sin A = \dfrac{\sqrt{17}}{17}$ and $\cot A = 4$.

> **Example 8**
> Given ABC is a triangle right angled at C such that $\cot A = \dfrac{1}{4}$. Find $\sin A, \cos A$ and $\tan A$.

Solution. Find $\sin A, \cos A$ and $\tan A$.
We know that $1 + \cot^2 A = \dfrac{1}{\sin^2 A}$.
It follows that $\sin^2 A = \dfrac{1}{1 + \cot^2 A}$.
By knowing that $\cot A = \dfrac{1}{4}$, we obtain
$$\sin^2 A = \frac{1}{1 + \left(\frac{1}{4}\right)^2}$$

$$= \frac{1}{1 + \frac{1}{16}}$$
$$= \frac{1}{\frac{17}{16}}$$
$$= \frac{16}{17}.$$

Then $\sin A = \frac{4}{\sqrt{17}} = \frac{4\sqrt{17}}{17}$.

Furthermore, $\sin^2 A + \cos^2 A = 1$.

It follows that
$$\cos^2 A = 1 - \sin^2 A$$
$$= 1 - \left(\frac{4}{\sqrt{17}}\right)^2$$
$$= 1 - \frac{16}{17}$$
$$= \frac{1}{17}.$$

It implies that $\cos A = \frac{1}{\sqrt{17}} = \frac{\sqrt{17}}{17}$.

Additionally, $\tan A = \frac{1}{\cot A} = \frac{1}{\frac{1}{4}} = 4$.

Therefore, $\sin A = \frac{4\sqrt{17}}{17}$, $\cos A = \frac{\sqrt{17}}{17}$ and $\tan A = 4$.

Chapter 1. Think in a Right Triangle

Chapter 2

The values of Trigonometric Functions of Special Angles

2.1 Trigonometric functions of angle 45°

Suppose that ABC is an isosceles right triangle(See the figure).

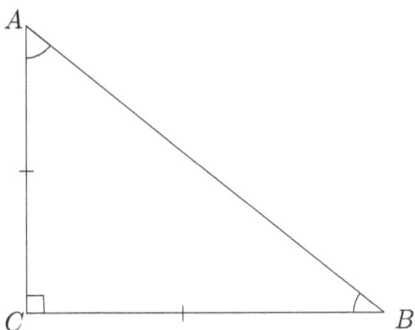

Remark 1. An Isosceles Right Triangle is a right triangle that consists of two equal length legs.

Chapter 2. The values of Trigonometric Functions of Special Angles

From the definitions, $\sin A = \dfrac{BC}{AB}$.
Using Pythagorean theorem, we obtain

$$\begin{aligned}AB^2 &= AC^2 + BC^2 \\ &= BC^2 + BC^2 \\ &= 2BC^2.\end{aligned}$$

Then $AB = \sqrt{2BC^2} = BC\sqrt{2}$.
It follows that $\sin A = \dfrac{BC}{AB} = \dfrac{BC}{BC\sqrt{2}} = \dfrac{1}{\sqrt{2}} = \dfrac{\sqrt{2}}{2}$.
Likewise,

$$\cos A = \dfrac{AC}{AB} = \dfrac{BC}{AB} = \dfrac{BC}{BC\sqrt{2}} = \dfrac{1}{\sqrt{2}} = \dfrac{\sqrt{2}}{2},$$

$$\tan A = \dfrac{BC}{AC} = \dfrac{AC}{AC} = 1,$$

and $\quad \cot A = \dfrac{AC}{BC} = \dfrac{AC}{AC} = 1.$

Since ABC is an isosceles right triangle, then $A = B = 45°$.
Therefore, $\quad \sin 45° = \dfrac{\sqrt{2}}{2}$

$$\cos 45° = \dfrac{\sqrt{2}}{2}$$

$$\tan 45° = 1$$

and $\quad \cot 45° = 1.$

2.2 Trigonometric functions of angles $30°$ and $60°$

Let ABC be a triangle right angled at C. Suppose that $\angle A = 30°$. Then $\angle B = 60°$ (the sum of all acute angles in a right triangle is equal to $90°$).

2.2. Trigonometric functions of angles 30° and 60°

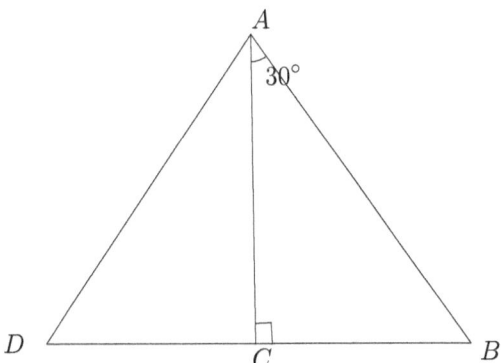

Let D be the reflection point of B over $[AC]$. Then ADB is an equilateral triangle. It follows that $BC = \dfrac{BD}{2} = \dfrac{AB}{2}$. Moreover, using Pythagorean theorem, we obtain

$$AC^2 = AB^2 - BC^2$$
$$= AB^2 - \left(\dfrac{AB}{2}\right)^2$$
$$= AB^2 - \dfrac{AB^2}{4}$$
$$= \dfrac{3AB^2}{4}.$$

Then $AC = \dfrac{AB\sqrt{3}}{2}$.

We have $\sin 30° = \dfrac{BC}{AB} = \dfrac{\frac{AB}{2}}{AB} = \dfrac{1}{2}$.
Additionally,

$$\cos 30° = \dfrac{AC}{AB} = \dfrac{\frac{\sqrt{3}}{2}AB}{AB} = \dfrac{\sqrt{3}}{2},$$

$$\tan 30° = \dfrac{\sin 30°}{\cos 30°} = \dfrac{\frac{1}{2}}{\frac{\sqrt{3}}{2}} = \dfrac{1}{2} \times \dfrac{2}{\sqrt{3}} = \dfrac{1}{\sqrt{3}} = \dfrac{\sqrt{3}}{3},$$

Chapter 2. The values of Trigonometric Functions of Special Angles

	$0°$	$30°$	$45°$	$60°$	$90°$
sin	0	$\dfrac{1}{2}$	$\dfrac{\sqrt{2}}{2}$	$\dfrac{\sqrt{3}}{2}$	1
cos	1	$\dfrac{\sqrt{3}}{2}$	$\dfrac{\sqrt{2}}{2}$	$\dfrac{1}{2}$	0
tan	0	$\dfrac{\sqrt{3}}{3}$	1	$\sqrt{3}$	∞
cot	∞	$\sqrt{3}$	1	$\dfrac{\sqrt{3}}{3}$	0

Figure 2.1: Trigonometric Functions of Special Angles

and $\quad \cot 30° = \dfrac{\cos 30°}{\sin 30°} = \dfrac{\frac{\sqrt{3}}{2}}{\frac{1}{2}} = \dfrac{\sqrt{3}}{2} \times 2 = \sqrt{3}.$

Likewise, $\sin 60° = \dfrac{\sqrt{3}}{2}$, $\cos 60° = \dfrac{1}{2}$, $\tan 60° = \sqrt{3}$ and $\cot 60° = \dfrac{\sqrt{3}}{3}.$

2.3 Trigonometric Functions of Special Angles

Figure 2.1 is the table of values of trigonometric functions of special angles.

Example 9

Calculate the following expressions

1. $A = \sin^2 30° - \cos 60° + \tan 45°$;
2. $B = \sin^2 45° - \tan^2 60° + \cot^2 45°.$

Solution.   1. $A = \sin^2 30° - \cos 60° + \tan 45°$

We have $A = \sin^2 30° - \cos 60° + \tan 45°$

2.3. Trigonometric Functions of Special Angles

$$= \left(\frac{1}{2}\right)^2 - \frac{1}{2} + 1$$
$$= \frac{1}{4} - \frac{1}{2} + 1$$
$$= \frac{1}{4} - \frac{2}{4} + \frac{4}{4}$$
$$= \frac{1-2+4}{4} = \frac{3}{4}.$$

Therefore, $A = \frac{3}{4}.$

2. $B = \sin^2 45° - \tan^2 60° + \cot^2 45°$

We have $B = \sin^2 45° - \tan^2 60° + \cot^2 45°$

$$= \left(\frac{\sqrt{2}}{2}\right)^2 - \sqrt{3}^2 + 1^2$$
$$= \frac{1}{2} - 3 + 1$$
$$= \frac{1}{2} - 2$$
$$= \frac{1}{2} - \frac{4}{2}$$
$$= \frac{1-4}{2} = -\frac{3}{2}.$$

Therefore, $B = -\frac{3}{2}.$

> **Example 10**
>
> Calculate the following expressions:
>
> 1. $A = \dfrac{\sin 30° + \cos 45°}{\tan 45° - \tan 60°}$;
> 2. $B = \tan 45° + \tan 30° \tan 60°.$

Solution. Calculate the expressions:

1. $A = \dfrac{\sin 30° + \cos 45°}{\tan 45° - \tan 60°}$

We have $A = \dfrac{\sin 30° + \cos 45°}{\tan 45° - \tan 60°}$

Chapter 2. The values of Trigonometric Functions of Special Angles

$$= \frac{\frac{1}{2} + \frac{\sqrt{2}}{2}}{1 - \sqrt{3}}$$

$$= \frac{\frac{1+\sqrt{2}}{2}}{1 - \sqrt{3}}$$

$$= \frac{1+\sqrt{2}}{2\left(1-\sqrt{3}\right)}$$

$$= \frac{\left(1+\sqrt{2}\right)\left(1+\sqrt{3}\right)}{2\left(1-\sqrt{3}\right)\left(1+\sqrt{3}\right)}$$

$$= \frac{1+\sqrt{3}+\sqrt{2}+\sqrt{6}}{2\left(1-\sqrt{3^2}\right)}$$

$$= \frac{1+\sqrt{2}+\sqrt{3}+\sqrt{6}}{-4}$$

$$= -\frac{1+\sqrt{2}+\sqrt{3}+\sqrt{6}}{4}$$

Consequently, $A = -\dfrac{1+\sqrt{2}+\sqrt{3}+\sqrt{6}}{4}$.

2. $B = \tan 45° + \tan 30° \tan 60°$

We have $B = \tan 45° + \tan 30° \tan 60°$

$$= 1 + \left(\frac{\sqrt{3}}{3}\right)\left(\sqrt{3}\right)$$

$$= 1 + \frac{\sqrt{3^2}}{3}$$

$$= 1 + 1 = 2.$$

Therefore, $B = 2$.

Chapter 3

Measure of Angles

We have already known that angle can be measured in degree. In this chapter, we will introduce readers another angular unit which is called Radian.

3.1 Radian

Radian is an angular unit that we use to describes the plane angle subtended by a circular arc. It is measured as the length of the arc divided by the radius of the circle

> **Definition 1**
> A radian is defined as the angle subtended at the center of a circle by an arc whose length is equal to the radius of the circle.

Basically, the magnitude in radians of such a subtended angle is equal to the ratio of the arc length to the radius of the circle; that is, $\theta = \frac{s}{r}$, where θ is the subtended angle in radians, s is the length of arc and r is the the radius of the circle. Notice that we use rad to denote radian in mathematics symbol. The radian symbol is mostly omitted. Using the fact that, the circumference of the circle is $C = 2\pi r$. It follows that the magnitude in radians of one complete revolution (360 degrees) is the length of the entire circumference divided by the radius, or $\frac{2\pi r}{r} = 2\pi$. Hence, $2\pi rad = 360°$. This is a very important relation for the conversion between radians and degrees.

3.2 Conversion between radians and degrees

From the above relation, it implies that

$$1° = \frac{\pi}{180} rad$$

and

$$1 rad = \frac{180°}{\pi}.$$

Remark 2. To be convenient, readers should remember that $\pi = 180°$.

> **Example 11**
>
> Convert from degrees to radians the given angles below:
>
> - $180° = 180 \times \dfrac{\pi}{180} = \pi$
> - $150° = 150 \times \dfrac{\pi}{180} = \dfrac{5\pi}{6}$
> - $135° = 135 \times \dfrac{\pi}{180} = \dfrac{3\pi}{4}$
> - $120° = 120 \times \dfrac{\pi}{180} = \dfrac{2\pi}{3}$
> - $90° = 90 \times \dfrac{\pi}{180} = \dfrac{\pi}{2}$
> - $60° = 60 \times \dfrac{\pi}{180} = \dfrac{\pi}{3}$
> - $45° = 45 \times \dfrac{\pi}{180} = \dfrac{\pi}{4}$
> - $30° = 30 \times \dfrac{\pi}{180} = \dfrac{\pi}{6}$

3.2. Conversion between radians and degrees

Example 12

Convert from radians to degrees the given angles below:

- $2\pi = 2\pi \times \dfrac{180°}{\pi} = 360°$

- $\pi = \pi \times \dfrac{180°}{\pi} = 180°$

- $\dfrac{\pi}{5} = \dfrac{\pi}{5} \times \dfrac{180°}{\pi} = 36°$

- $\dfrac{\pi}{8} = \dfrac{\pi}{8} \times \dfrac{180°}{\pi} = 22.5°$

- $\dfrac{\pi}{10} = \dfrac{\pi}{10} \times \dfrac{180°}{\pi} = 18°$

- $\dfrac{\pi}{12} = \dfrac{\pi}{12} \times \dfrac{180°}{\pi} = 15°$

It is easier to convert angles from radians to degrees by just substitute π by $180°$. For example, $\dfrac{5\pi}{6} = \dfrac{5}{6} \times 180° = 150°$.

Example 13

Convert from radians to degrees the given angles below:

- $\dfrac{2\pi}{3} = \dfrac{2}{3} \times 180° = 120°$

- $\dfrac{4\pi}{3} = \dfrac{4}{3} \times 180° = 240°$

- $\dfrac{3\pi}{2} = \dfrac{3}{2} \times 180° = 270°$

- $\dfrac{7\pi}{6} = \dfrac{7}{6} \times 180° = 210°$

- $\dfrac{4\pi}{5} = \dfrac{4}{5} \times 180° = 144°$

- $\dfrac{5\pi}{6} = \dfrac{5}{6} \times 180° = 150°$

Chapter 3. Measure of Angles

Chapter 4

Trigonometric Functions in the Unit Circle

We have already known about trigonometric functions in a right triangle. In this chapter, we will generalize the definition of trigonometric functions in a more broader way. That is, we will study the concept of trigonometric functions in the unit circle. We start this chapter by given the definition of the unit circle.

> **Definition 2**
> The unit circle is the circle of radius one centered at the origin (see Figure 4.1).

In the unit circle, we defined angle by two vectors $\overrightarrow{OP_0}$ and $\overrightarrow{OP}$, where P_0 is the intersection point of the unit circle and the positive part of x-axis and P_1 is an arbitrary point on the circle. An angle is formed when we rotate $\overrightarrow{OP}$ from $\overrightarrow{OP_0}$ (see the figure). When $\overrightarrow{OP}$ is rotated in the clockwise direction, we obtain a negative angle. Conversely, we obtain a positive angle if $\overrightarrow{OP}$ is rotated in the anti-clockwise direction. Notice that when $\overrightarrow{OP_0}$ and $\overrightarrow{OP}$ forms angle θ, it is generally written as

$$\left(\overrightarrow{OP_0}, \overrightarrow{OP}\right) = \theta.$$

Chapter 4. Trigonometric Functions in the Unit Circle

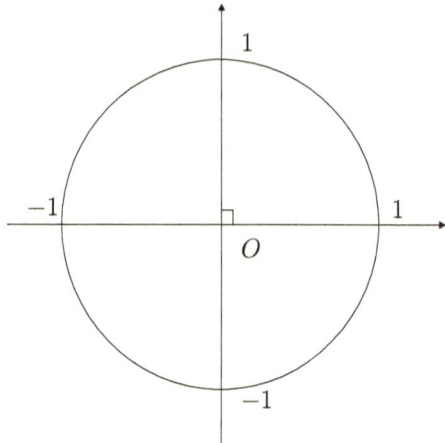

Figure 4.1: Unit Circle

Let P_0 and $P(x, y)$ be a point on the unit circle such that $\left(\overrightarrow{OP_0}, \overrightarrow{OP}\right) = \theta$. Let H be a perpendicular foot of P on the horizontal axis (See Figure 4.2). In right triangle POH, we obtain

$$\sin \angle POH = \frac{PH}{OP} = \frac{y}{1} = y,$$
$$\cos \angle POH = \frac{OH}{OP} = \frac{x}{1} = x,$$
$$\tan \angle POH = \frac{PH}{OH} = \frac{y}{x},$$
$$\text{and} \quad \cot \angle POH = \frac{OH}{PH} = \frac{x}{y}.$$

Since $\angle POH = \left(\overrightarrow{OP_0}, \overrightarrow{OP}\right) = \theta$, it follows that

$$\sin \theta = y$$
$$\cos \theta = x$$
$$\tan \theta = \frac{y}{x}$$

24

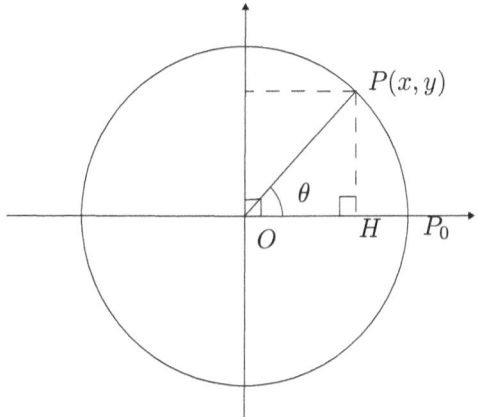

Figure 4.2: P is an arbitrary point on the circle

$$\text{and} \quad \cot\theta = \frac{x}{y}.$$

Generally, suppose P_0 and $P(x,y)$ is a point on the unit circle. Taking $\left(\overrightarrow{OP_0}, \overrightarrow{OP}\right) = \theta$, then $\sin\theta = y$, $\cos\theta = x$, $\tan\theta = \frac{y}{x}$ and $\cot\theta = \frac{x}{y}$. That is $P(\cos\theta, \sin\theta)$. From the above observation, to extend the concept of trigonometric functions in right triangle to the unit circle, we defined trigonometric functions in the unit circle as below:

General 1

Given that $\left(\overrightarrow{OP_0}, \overrightarrow{OP}\right) = \theta$, where $P(x,y)$. We defined

$$\sin\theta = y$$
$$\cos\theta = x$$
$$\tan\theta = \frac{y}{x}$$
$$\text{and} \quad \cot\theta = \frac{x}{y}.$$

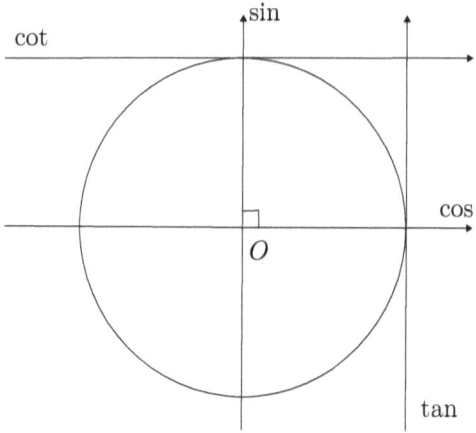

Figure 4.3: Trigonometric Functions in the Unit Circle

4.1 Symmetry, shifts, and periodicity

By examining the unit circle, the following properties of the trigonometric functions can be established.

4.1.1 Symmetry

Given that α and β are two angles reflected in angle γ. The trigonometric functions of α are often one of the other trigonometric functions of β. This leads to the following identities:

1. Reflected in 0
   We obtain the following formulas:
   $$\sin(-\theta) = -\sin\theta$$
   $$\cos(-\theta) = \cos\theta$$
   $$\tan(-\theta) = -\cot\theta$$
   $$\cot(-\theta) = -\tan\theta$$

2. Reflected in $\dfrac{\pi}{4}$ (Co-function identities)
   $$\sin\left(\frac{\pi}{2} - \theta\right) = \cos\theta$$

4.1. Symmetry, shifts, and periodicity

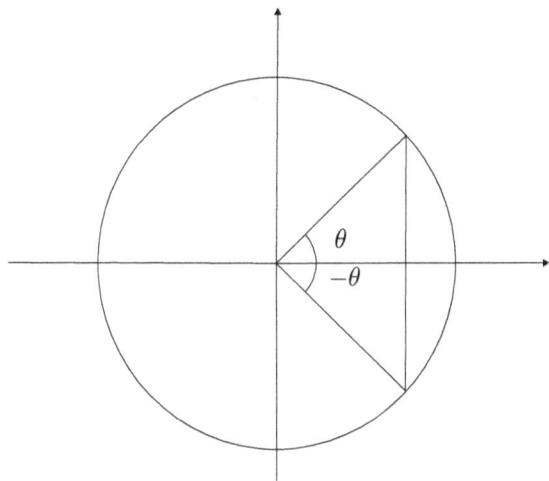

Figure 4.4: Reflected in 0

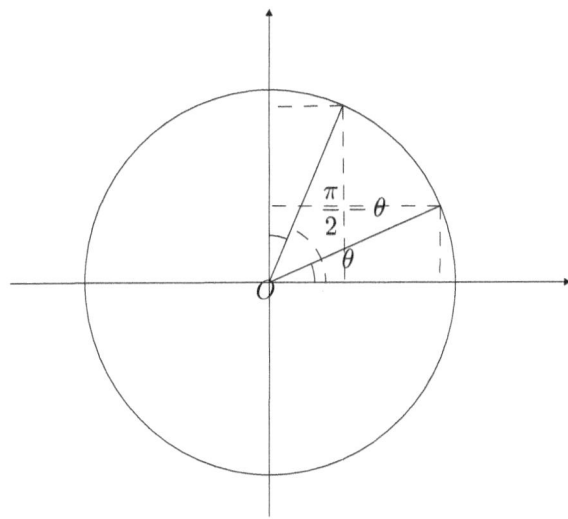

Figure 4.5: Reflected in $\dfrac{\pi}{4}$

Chapter 4. Trigonometric Functions in the Unit Circle

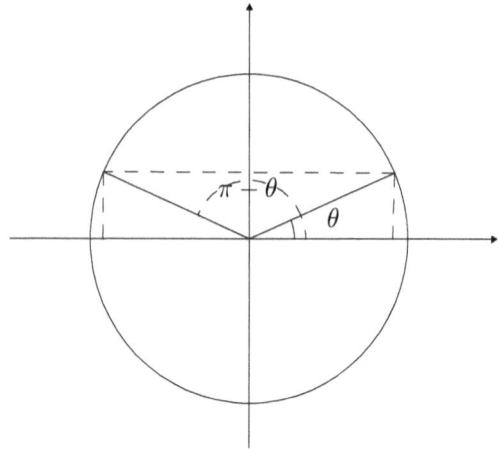

Figure 4.6: Reflected in $\frac{\pi}{2}$

$$\cos\left(\frac{\pi}{2} - \theta\right) = \sin\theta$$
$$\tan\left(\frac{\pi}{2} - \theta\right) = \cot\theta$$
$$\cot\left(\frac{\pi}{2} - \theta\right) = \tan\theta$$

3. Reflected in $\frac{\pi}{2}$.

$$\sin(\pi - \theta) = \sin\theta$$
$$\cos(\pi - \theta) = -\cos\theta$$
$$\tan(\pi - \theta) = -\tan\theta$$
$$\cot(\pi - \theta) = -\cot\theta$$

4.1.2 Shifts and periodicity

1. Shift by $\frac{\pi}{2}$

Using reflected in $\frac{\pi}{4}$ and 0 formulas, the following formulas

4.1. Symmetry, shifts, and periodicity

yield.

$$\sin\left(\frac{\pi}{2} + \theta\right) = \cos\theta$$
$$\cos\left(\frac{\pi}{2} + \theta\right) = -\sin\theta$$
$$\tan\left(\frac{\pi}{2} + \theta\right) = -\tan\theta$$
$$\cot\left(\frac{\pi}{2} + \theta\right) = -\cot\theta$$

2. Shift by π (Period for $\tan x$ and $\cot x$)
   Using reflected in $\frac{\pi}{2}$ and 0 formulas, the following formulas yield.

$$\sin(\pi + \theta) = -\sin\theta$$
$$\cos(\pi + \theta) = -\cos\theta$$
$$\tan(\pi + \theta) = \tan\theta$$
$$\cot(\pi + \theta) = \cot\theta$$

3. θ and $\theta + 2k\pi$, where $k \in \mathbb{Z}$.

$$\sin(\theta + 2k\pi) = \sin\theta$$
$$\cos(\theta + 2k\pi) = \cos\theta$$
$$\tan(\theta + 2k\pi) = \tan\theta$$
$$\cot(\theta + 2k\pi) = \cot\theta$$

Chapter 4. Trigonometric Functions in the Unit Circle

Chapter 5

Trigonometric Identities

5.1 Addition and Difference Formulas

5.1.1 $\cos(\alpha - \beta)$

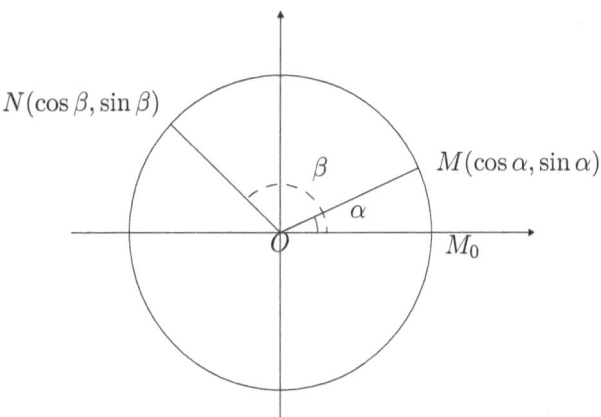

Let M_0, M and N be points on the unit circle such that $\angle MOM_0 = \alpha$ and $\angle NOM_0 = \beta$. Then $M(\cos\alpha, \sin\alpha)$ and $N(\cos\beta, \sin\beta)$. From the definition of scalar product,

$$\overrightarrow{OM} \cdot \overrightarrow{ON} = \left|\overrightarrow{OM}\right| \times \left|\overrightarrow{ON}\right| \cos(\alpha - \beta) = 1 \times 1 \times \cos(\alpha - \beta)$$

Chapter 5. Trigonometric Identities

or
$$\vec{OM} \cdot \vec{ON} = \cos(\alpha - \beta). \tag{1}$$

Moreover, $\vec{OM} = (\cos\alpha, \sin\alpha)$ and $\vec{ON} = (\cos\beta, \sin\beta)$.
It follows that
$$\vec{OM} \cdot \vec{ON} = \cos\alpha\cos\beta + \sin\alpha\sin\beta. \tag{2}$$

From (1) and (2), we obtain $\cos(\alpha - \beta) = \cos\alpha\cos\beta + \sin\alpha\sin\beta$.

> **General 2**
> For all angles α and β, we obtain
> $$\cos(\alpha - \beta) = \cos\alpha\cos\beta + \sin\alpha\sin\beta.$$

5.1.2 $\cos(\alpha + \beta)$

We have
$$\begin{aligned}\cos(\alpha + \beta) &= \cos[\alpha - (-\beta)] \\ &= \cos\alpha\cos(-\beta) + \sin\alpha\sin(-\beta) \\ &= \cos\alpha\cos\beta - \sin\alpha\sin\beta.\end{aligned}$$

Therefore, $\cos(\alpha + \beta) = \cos\alpha\cos\beta - \sin\alpha\sin\beta$.

> **General 3**
> For all angles α and β, we obtain
> $$\cos(\alpha + \beta) = \cos\alpha\cos\beta - \sin\alpha\sin\beta.$$

> **Example 14**
> Find the exact value of $\cos 15°$ and $\cos 75°$.

Solution. Using the formula, $\cos(\alpha - \beta) = \cos\alpha\cos\beta + \sin\alpha\sin\beta$.
We obtain
$$\begin{aligned}\cos 15° &= \cos(45° - 30°) \\ &= \cos 45° \cos 30° + \sin 45° \sin 30°\end{aligned}$$

5.1. Addition and Difference Formulas

$$= \left(\frac{\sqrt{2}}{2}\right)\left(\frac{\sqrt{3}}{2}\right) + \left(\frac{\sqrt{2}}{2}\right)\left(\frac{1}{2}\right)$$

$$= \frac{\sqrt{6}+\sqrt{2}}{4}$$

and

$$\cos 75° = \cos(45° + 30°)$$
$$= \cos 45° \cos 30° - \sin 45° \sin 30°$$
$$= \left(\frac{\sqrt{2}}{2}\right)\left(\frac{\sqrt{3}}{2}\right) - \left(\frac{\sqrt{2}}{2}\right)\left(\frac{1}{2}\right)$$
$$= \frac{\sqrt{6}-\sqrt{2}}{4}.$$

Therefore, $\cos 15° = \dfrac{\sqrt{6}+\sqrt{2}}{4}$ and $\cos 75° = \dfrac{\sqrt{6}-\sqrt{2}}{4}$.

5.1.3 $\sin(\alpha - \beta)$

We have
$$\sin(\alpha - \beta) = \cos\left[\frac{\pi}{2} - (\alpha - \beta)\right]$$
$$= \cos\left[\left(\frac{\pi}{2} - \alpha\right) + \beta\right]$$
$$= \cos\left(\frac{\pi}{2} - \alpha\right)\cos\beta - \sin\left(\frac{\pi}{2} - \alpha\right)\sin\beta$$
$$= \sin\alpha \cos\beta - \sin\beta \cos\alpha.$$

Therefore, $\sin(\alpha - \beta) = \sin\alpha \cos\beta - \sin\beta \cos\alpha$.

> **General 4**
>
> For all angles α and β, we obtain
>
> $$\sin(\alpha - \beta) = \sin\alpha \cos\beta - \sin\beta \cos\alpha.$$

> **Example 15**
>
> Find the exact value of $\sin 15°$.

Solution. Using $\sin(\alpha - \beta) = \sin\alpha\cos\beta - \sin\beta\cos\alpha$, it implies that

$$\sin 15° = \sin(45° - 30°)$$
$$= \sin 45° \cos 30° - \sin 30° \cos 45°$$
$$= \left(\frac{\sqrt{2}}{2}\right)\left(\frac{\sqrt{3}}{2}\right) - \left(\frac{1}{2}\right)\left(\frac{\sqrt{2}}{2}\right)$$
$$= \frac{\sqrt{6} - \sqrt{2}}{4}.$$

Therefore, $\sin 15° = \dfrac{\sqrt{6} - \sqrt{2}}{4}$.

5.1.4 $\sin(\alpha + \beta)$

We have
$$\sin(\alpha + \beta) = \sin[\alpha - (-\beta)]$$
$$= \sin\alpha\cos(-\beta) - \sin(-\beta)\cos\alpha$$
$$= \sin\alpha\cos\beta + \sin\beta\cos\alpha.$$

Therefore, $\sin(\alpha + \beta) = \sin\alpha\cos\beta + \sin\beta\cos\alpha$.

General 5

For all angles α and β, we obtain

$$\sin(\alpha + \beta) = \sin\alpha\cos\beta + \sin\beta\cos\alpha.$$

Example 16

Find the exact value of $\sin 75°$.

Solution. We have $\sin(\alpha + \beta) = \sin\alpha\cos\beta + \sin\beta\cos\alpha$. Then

$$\sin 75° = \sin(30° + 45°)$$
$$= \sin 30° \cos 45° + \sin 45° \cos 30°$$
$$= \left(\frac{1}{2}\right)\left(\frac{\sqrt{2}}{2}\right) + \left(\frac{\sqrt{2}}{2}\right)\left(\frac{\sqrt{3}}{2}\right)$$

5.1. Addition and Difference Formulas

$$= \frac{\sqrt{2} + \sqrt{6}}{4}.$$

Therefore, $\sin 75° = \dfrac{\sqrt{6} + \sqrt{2}}{4}$.

5.1.5 $\tan(\alpha - \beta)$

We have
$$\tan(\alpha - \beta) = \frac{\sin(\alpha - \beta)}{\cos(\alpha - \beta)}$$
$$= \frac{\sin\alpha\cos\beta - \sin\beta\cos\alpha}{\cos\alpha\cos\beta + \sin\alpha\sin\beta}$$
$$= \frac{\dfrac{\sin\alpha\cos\beta - \sin\beta\cos\alpha}{\cos\alpha\cos\beta}}{\dfrac{\cos\alpha\cos\beta + \sin\alpha\sin\beta}{\cos\alpha\cos\beta}}$$
$$= \frac{\dfrac{\sin\alpha\cos\beta}{\cos\alpha\cos\beta} - \dfrac{\sin\beta\cos\alpha}{\cos\alpha\cos\beta}}{\dfrac{\cos\alpha\cos\beta}{\cos\alpha\cos\beta} + \dfrac{\sin\alpha\sin\beta}{\cos\alpha\cos\beta}}$$
$$= \frac{\dfrac{\sin\alpha}{\cos\alpha} - \dfrac{\sin\beta}{\cos\beta}}{1 + \left(\dfrac{\sin\alpha}{\cos\alpha}\right)\left(\dfrac{\sin\beta}{\cos\beta}\right)}$$
$$= \frac{\tan\alpha - \tan\beta}{1 + \tan\alpha\tan\beta}.$$

Therefore, $\tan(\alpha - \beta) = \dfrac{\tan\alpha - \tan\beta}{1 + \tan\alpha\tan\beta}$.

General 6

For all angles α and β such that $\alpha, \beta \neq \dfrac{\pi}{2} + k\pi$ and $\alpha - \beta \neq \dfrac{\pi}{2} + k\pi$, we obtain

$$\tan(\alpha - \beta) = \frac{\tan\alpha - \tan\beta}{1 + \tan\alpha\tan\beta}.$$

> **Example 17**
> Find the exact value of tan 15°.

Solution. We have

$$\tan 15° = \tan(45° - 30°)$$
$$= \frac{\tan 45° - \tan 30°}{1 + \tan 45° \tan 30°}$$
$$= \frac{1 - \frac{\sqrt{3}}{3}}{1 + (1)\left(\frac{\sqrt{3}}{3}\right)}$$
$$= \frac{\frac{3 - \sqrt{3}}{3}}{\frac{3 + \sqrt{3}}{3}}$$
$$= \frac{3 - \sqrt{3}}{3 + \sqrt{3}} = \frac{\sqrt{3} - 1}{\sqrt{3} + 1}$$
$$= \frac{\left(\sqrt{3} - 1\right)^2}{\left(\sqrt{3} + 1\right)\left(\sqrt{3} - 1\right)}$$
$$= \frac{3 - 2\sqrt{3} + 1}{3 - 1}$$
$$= \frac{4 - 2\sqrt{3}}{2}$$
$$= \frac{2\left(2 - \sqrt{3}\right)}{2}$$
$$= 2 - \sqrt{3}.$$

Therefore, $\tan 15° = 2 - \sqrt{3}$.

5.1.6 $\tan(\alpha + \beta)$

We have $\tan(\alpha + \beta) = \tan[\alpha - (-\beta)]$
$$= \frac{\tan \alpha - \tan(-\beta)}{1 + \tan \alpha \tan(-\beta)}$$

5.1. Addition and Difference Formulas

$$= \frac{\tan\alpha + \tan\beta}{1 - \tan\alpha\tan\beta}.$$

Therefore, $\tan(\alpha + \beta) = \dfrac{\tan\alpha + \tan\beta}{1 - \tan\alpha\tan\beta}.$

> **General 7**
>
> For all angles α and β such that $\alpha, \beta \neq \dfrac{\pi}{2} + k\pi$ and $\alpha + \beta \neq \dfrac{\pi}{2} + k\pi$, we obtain
>
> $$\tan(\alpha + \beta) = \frac{\tan\alpha + \tan\beta}{1 - \tan\alpha\tan\beta}.$$

> **Example 18**
>
> Find the exact value of $\tan 75°$.

Solution.

We have
$$\tan 75° = \tan(30° + 45°)$$
$$= \frac{\tan 30° + \tan 45°}{1 - \tan 30° \tan 45°}$$
$$= \frac{\dfrac{\sqrt{3}}{3} + 1}{1 - \dfrac{\sqrt{3}}{3}}$$
$$= \frac{\dfrac{\sqrt{3}+3}{3}}{\dfrac{3-\sqrt{3}}{3}}$$
$$= \frac{\sqrt{3}+3}{3-\sqrt{3}} = \frac{\sqrt{3}+1}{\sqrt{3}-1}$$
$$= \frac{(\sqrt{3}+1)^2}{(\sqrt{3}-1)(\sqrt{3}+1)}$$
$$= \frac{3 + 2\sqrt{3} + 1}{3 - 1}$$
$$= \frac{4 + 2\sqrt{3}}{2}$$

$$= \frac{2\left(2+\sqrt{3}\right)}{2}$$
$$= 2+\sqrt{3}.$$

Therefore, $\tan 75° = 2 + \sqrt{3}$.

Notice that $\cot x$ is the reciprocal of $\tan x$. So, we do not mainly discuss the formula for $\cot x$. Anyway, the readers should find it by themselves.

5.2 Double-Angle Formulas

We have
$$\sin 2\alpha = \sin(\alpha + \alpha)$$
$$= \sin\alpha\cos\alpha + \sin\alpha\cos\alpha$$
$$= 2\sin\alpha\cos\alpha.$$

Moreover, $\cos 2\alpha = \cos(\alpha + \alpha)$
$$= \cos\alpha\cos\alpha - \sin\alpha\sin\alpha$$
$$= \cos^2\alpha - \sin^2\alpha.$$

By knowing that $\sin^2\alpha + \cos^2\alpha = 1$ or $\cos^2\alpha = 1 - \sin^2\alpha$, then
$$\cos 2\alpha = 1 - \sin^2\alpha - \sin^2\alpha = 1 - 2\sin^2\alpha.$$

Additionally,
$$\cos 2\alpha = \cos^2\alpha - \left(1 - \cos^2\alpha\right)$$
$$= \cos^2\alpha - 1 + \cos^2\alpha$$
$$= 2\cos^2\alpha - 1.$$

Another important double-angle formula for solving trigonometric problems is the formula for $\tan 2\alpha$.

Observe that $\tan 2\alpha = \tan(\alpha + \alpha) = \dfrac{\tan\alpha + \tan\alpha}{1 - \tan\alpha\tan\alpha} = \dfrac{2\tan\alpha}{1 - \tan^2\alpha}$.

To sum up,
$$\sin 2\alpha = 2\sin\alpha\cos\alpha$$

5.2. Double-Angle Formulas

$$\cos 2\alpha = \cos^2\alpha - \sin^2\alpha = 1 - 2\sin^2\alpha = 2\cos^2\alpha - 1$$

and $\quad \tan 2\alpha = \dfrac{2\tan\alpha}{1-\tan^2\alpha}.$

Example 19

Express $\sin 3x$, $\cos 3x$ and $\tan 3x$ in terms of $\sin x$, $\cos x$ and $\tan x$ respectively.

Solution. We have

$$\begin{aligned}
\sin 3x &= \sin(2x + x) \\
&= \sin 2x \cos x + \sin x \cos 2x \\
&= 2\sin x \cos x \cos x + \sin x \left(1 - 2\sin^2 x\right) \\
&= 2\sin x \cos^2 x + \sin x - 2\sin^3 x \\
&= 2\sin x \left(1 - \sin^2 x\right) + \sin x - 2\sin^3 x \\
&= 2\sin x - 2\sin^3 x + \sin x - 2\sin^3 x \\
&= 3\sin x - 4\sin^3 x
\end{aligned}$$

and

$$\begin{aligned}
\cos 3x &= \cos(2x + x) \\
&= \cos 2x \cos x - \sin 2x \sin x \\
&= \left(2\cos^2 x - 1\right)\cos x - 2\sin x \cos x \sin x \\
&= 2\cos^3 x - \cos x - 2\sin^2 x \cos x \\
&= 2\cos^3 x - \cos x - 2\left(1 - \cos^2 x\right)\cos x \\
&= 2\cos^3 x - \cos x - 2\cos x + 2\cos^2 x \\
&= 4\cos^3 x - 3\cos x.
\end{aligned}$$

Additionally,

$$\begin{aligned}
\tan 3x = \tan(2x + x) &= \dfrac{\tan 2x + \tan x}{1 - \tan 2x \tan x} \\
&= \dfrac{\dfrac{2\tan x}{1-\tan^2 x} + \tan x}{1 - \tan x \left(\dfrac{2\tan x}{1-\tan^2 x}\right)}
\end{aligned}$$

Chapter 5. Trigonometric Identities

$$= \frac{\frac{2\tan x + \tan x - \tan^3 x}{1 - \tan^2 x}}{\frac{1 - \tan^2 - 2\tan^2 x}{1 - \tan^2 x}}$$

$$= \frac{3\tan x - \tan^3 x}{1 - 3\tan^2 x}.$$

> **Example 20**
>
> Given that $\sin x = \frac{3}{4}$ and $\frac{\pi}{2} < x < \pi$. Find the values of $\cos 2x$, $\sin 2x$ and $\tan 2x$.

Solution. • Find $\cos 2x$.
Using $\cos 2x = 1 - 2\sin^2 x$, we obtain

$$\cos 2x = 1 - 2\sin^2 x$$
$$= 1 - 2\left(\frac{3}{4}\right)^2$$
$$= 1 - 2\left(\frac{9}{16}\right)$$
$$= 1 - \frac{9}{8}$$
$$= -\frac{1}{8}.$$

Therefore, $\cos 2x = -\frac{1}{8}$.

• Find $\sin 2x$.
We have $\sin^2 x + \cos^2 x = 1$.
Then $\cos^2 x = 1 - \sin^2 x = 1 - \left(\frac{3}{4}\right)^2 = 1 - \frac{9}{16} = \frac{7}{16}$.
It follows that $\cos x = -\frac{\sqrt{7}}{4}$ since $\frac{\pi}{2} < x < \pi$.
We obtain

$$\sin 2x = 2\sin x \cos x$$
$$= 2 \times \frac{3}{4} \times \left(-\frac{\sqrt{7}}{4}\right)$$

5.2. Double-Angle Formulas

$$= -\frac{3\sqrt{7}}{8}$$

Therefore, $\sin 2x = -\dfrac{3\sqrt{7}}{8}$.

- Find $\tan 2x$.

  We have $\tan 2x = \dfrac{\sin 2x}{\cos 2x} = \dfrac{-\dfrac{3\sqrt{7}}{8}}{-\dfrac{1}{8}} = 3\sqrt{7}$.

  Therefore, $\tan 2x = 3\sqrt{7}$.

> **Example 21**
> Prove that $\dfrac{1 - 2\sin^2 x}{2} = \cot\left(\dfrac{\pi}{4} + x\right)\cos^2\left(\dfrac{\pi}{4} - x\right)$.

Solution. We have

$$\frac{1 - 2\sin^2 x}{2\cot\left(\dfrac{\pi}{4} + x\right)\cos^2\left(\dfrac{\pi}{4} - x\right)} = \frac{\cos 2x}{2\cot\left(\dfrac{\pi}{4} + x\right)\sin^2\left[\dfrac{\pi}{2} - \left(\dfrac{\pi}{4} - x\right)\right]}$$

$$= \frac{\cos 2x}{2\cot\left(\dfrac{\pi}{4} + x\right)\sin^2\left(\dfrac{\pi}{4} + x\right)}$$

$$= \frac{\cos 2x}{\dfrac{2\cos\left(\dfrac{\pi}{4} + x\right)}{\sin\left(\dfrac{\pi}{4} + x\right)} \times \sin^2\left(\dfrac{\pi}{4} + x\right)}$$

$$= \frac{\cos 2x}{2\cos\left(\dfrac{\pi}{4} + x\right)\sin\left(\dfrac{\pi}{4} + x\right)}$$

$$= \frac{\cos 2x}{\sin 2\left(\dfrac{\pi}{4} + x\right)}$$

$$= \frac{\cos 2x}{\sin\left(\dfrac{\pi}{2} + 2x\right)}$$

$$= \frac{\cos 2x}{\cos 2x}$$

$$= 1.$$

Therefore, $\dfrac{1-2\sin^2 x}{2} = \cot\left(\dfrac{\pi}{4}+x\right)\cos^2\left(\dfrac{\pi}{4}-x\right)$.

5.3 Half-Angle Formulas

> **General 8**
>
> Generally, $\sin^2\dfrac{x}{2} = \dfrac{1-\cos x}{2}$, $\cos^2\dfrac{x}{2} = \dfrac{1+\cos x}{2}$ and $\tan^2\dfrac{x}{2} = \dfrac{1-\cos x}{1+\cos x}$.

Proof. We have $\cos x = \cos 2\left(\dfrac{x}{2}\right) = 1 - 2\sin^2\dfrac{x}{2}$.
Then $\sin^2\dfrac{x}{2} = \dfrac{1-\cos x}{2}$.
Additionally, $\cos x = \cos 2\left(\dfrac{x}{2}\right) = 2\cos^2\dfrac{x}{2} - 1$.
It follows that $\cos^2\dfrac{x}{2} = \dfrac{1+\cos x}{2}$.
Moreover, $\tan^2\dfrac{x}{2} = \dfrac{\sin^2\dfrac{x}{2}}{\cos^2\dfrac{x}{2}} = \dfrac{\dfrac{1-\cos x}{2}}{\dfrac{1+\cos x}{2}} = \dfrac{1-\cos x}{1+\cos x}$.
Thus, the given formulas are proved. □

> **Example 22**
>
> Find the exact values of $\sin\dfrac{\pi}{8}$, $\cos\dfrac{\pi}{8}$ and $\tan\dfrac{\pi}{8}$.

Solution. • Find $\sin\dfrac{\pi}{8}$.

We have $\sin^2\dfrac{\pi}{8} = \dfrac{1-\cos\dfrac{\pi}{4}}{2} = \dfrac{1-\dfrac{\sqrt{2}}{2}}{2} = \dfrac{\dfrac{2-\sqrt{2}}{2}}{2} = \dfrac{2-\sqrt{2}}{4}$.

Since $\sin\dfrac{\pi}{8} > 0$, we obtain $\sin\dfrac{\pi}{8} = \dfrac{\sqrt{2-\sqrt{2}}}{2}$.

Therefore, $\sin\dfrac{\pi}{8} = \dfrac{\sqrt{2-\sqrt{2}}}{2}$.

• Find $\cos\dfrac{\pi}{8}$.

5.3. Half-Angle Formulas

We have $\cos^2 \dfrac{\pi}{8} = \dfrac{1+\cos \dfrac{\pi}{4}}{2} = \dfrac{1+\dfrac{\sqrt{2}}{2}}{2} = \dfrac{\dfrac{2+\sqrt{2}}{2}}{2} = \dfrac{2+\sqrt{2}}{4}$.

Since $\cos \dfrac{\pi}{8} > 0$, it follows that $\cos \dfrac{\pi}{8} = \dfrac{\sqrt{2+\sqrt{2}}}{2}$.

Therefore, $\cos \dfrac{\pi}{8} = \dfrac{\sqrt{2+\sqrt{2}}}{2}$.

- Find $\tan \dfrac{\pi}{8}$.

We have $\tan^2 \dfrac{\pi}{8} = \dfrac{1-\cos \dfrac{\pi}{8}}{1+\cos \dfrac{\pi}{8}}$

$= \dfrac{1-\dfrac{\sqrt{2}}{2}}{1+\dfrac{\sqrt{2}}{2}}$

$= \dfrac{\dfrac{2-\sqrt{2}}{2}}{\dfrac{2+\sqrt{2}}{2}}$

$= \dfrac{2-\sqrt{2}}{2+\sqrt{2}}$

$= \dfrac{\left(2-\sqrt{2}\right)^2}{\left(2+\sqrt{2}\right)\left(2-\sqrt{2}\right)}$

$= \dfrac{4-4\sqrt{2}+2}{4-2}$

$= \dfrac{6-4\sqrt{2}}{2}$

$= \dfrac{2\left(3-2\sqrt{2}\right)}{2}$

$= 3-2\sqrt{2}$.

By knowing that $\tan \dfrac{\pi}{8} > 0$, it follows that $\tan \dfrac{\pi}{8} = \sqrt{3-2\sqrt{2}}$.

Therefore, $\tan \dfrac{\pi}{8} = \sqrt{3-2\sqrt{2}}$.

5.4 $\sin x$, $\cos x$ and $\tan x$ in terms of $t = \tan \dfrac{x}{2}$

> **General 9**
>
> Suppose that $t = \tan \dfrac{x}{2}$. We obtain $\cos x = \dfrac{1-t^2}{1+t^2}$, $\sin x = \dfrac{2t}{1+t^2}$ and $\tan x = \dfrac{2t}{1-t^2}$.

Proof.

We have
$$\cos x = \cos 2\left(\frac{x}{2}\right)$$
$$= \cos^2 \frac{x}{2} - \sin^2 \frac{x}{2}$$
$$= \frac{\cos^2 \dfrac{x}{2} - \sin^2 \dfrac{x}{2}}{\cos^2 \dfrac{x}{2} + \sin^2 \dfrac{x}{2}}$$
$$= \frac{\dfrac{\cos^2 \dfrac{x}{2} - \sin^2 \dfrac{x}{2}}{\cos^2 \dfrac{x}{2}}}{\dfrac{\cos^2 \dfrac{x}{2} + \sin^2 \dfrac{x}{2}}{\cos^2 \dfrac{x}{2}}}$$
$$= \frac{1 - \left(\dfrac{\sin \dfrac{x}{2}}{\cos \dfrac{x}{2}}\right)^2}{1 + \left(\dfrac{\sin \dfrac{x}{2}}{\cos \dfrac{x}{2}}\right)^2}$$
$$= \frac{1 - t^2}{1 + t^2}.$$

5.4. $\sin x$, $\cos x$ and $\tan x$ in terms of $t = \tan \dfrac{x}{2}$

In addition,

$$\sin x = \sin\left(\frac{x}{2}\right) = 2\sin\frac{x}{2}\cos\frac{x}{2} = \frac{2\sin\dfrac{x}{2}\cos\dfrac{x}{2}}{\sin^2\dfrac{x}{2}+\cos^2\dfrac{x}{2}}$$

$$= \frac{\dfrac{2\sin\dfrac{x}{2}\cos\dfrac{x}{2}}{\cos^2\dfrac{x}{2}}}{\dfrac{\sin^2\dfrac{x}{2}+\cos^2\dfrac{x}{2}}{\cos^2\dfrac{x}{2}}}$$

$$= \frac{\dfrac{2\sin\dfrac{x}{2}}{\cos\dfrac{x}{2}}}{\left(\dfrac{\sin\dfrac{x}{2}}{\cos\dfrac{x}{2}}\right)^2+1}$$

$$= \frac{2t}{1+t^2}.$$

Finally, $\tan x = \dfrac{2\tan\dfrac{x}{2}}{1-\tan^2\dfrac{x}{2}} = \dfrac{2t}{1-t^2}.$ ☐

Example 23

Prove that $\dfrac{1-\cos x}{\sin x} = \tan\dfrac{x}{2}.$

Solution. By knowing that $\cos x = \dfrac{1-t^2}{1+t^2}$ and $\sin x = \dfrac{2t}{1+t^2}$, where $t = \tan\dfrac{x}{2}$, it follows that

$$\frac{1-\cos x}{\sin x} = \frac{1-\dfrac{1-t^2}{1+t^2}}{\dfrac{2t}{1+t^2}}$$

45

$$= \frac{1+t^2-1+t^2}{\frac{1+t^2}{\frac{2t}{1+t^2}}}$$

$$= \left(\frac{2t^2}{1+t^2}\right)\left(\frac{1+t^2}{2t}\right)$$

$$= t = \tan\frac{x}{2}.$$

5.5 Product to Sum Formulas

> **General 10**
> - $\cos a \cos b = \dfrac{1}{2}[\cos(a-b) + \cos(a+b)]$;
> - $\sin a \sin b = \dfrac{1}{2}[\cos(a-b) - \cos(a+b)]$;
> - $\sin a \cos b = \dfrac{1}{2}[\sin(a+b) + \sin(a-b)]$;
> - $\sin b \cos a = \dfrac{1}{2}[\sin(a+b) - \sin(a-b)]$.

Proof. We have

$$\cos(a-b) = \cos a \cos b + \sin a \sin b. \qquad (1)$$

and

$$\cos(a+b) = \cos a \cos b - \sin a \sin b. \qquad (2)$$

Adding (1) and (2), we obtain

$$\cos(a-b) + \cos(a+b) = 2\cos a \cos b$$

or

$$\cos a \cos b = \frac{1}{2}[\cos(a-b) + \cos(a+b)].$$

Subtract (1) by (2), it follows that

$$\cos(a-b) - \cos(a+b) = 2\sin a \sin b$$

5.5. Product to Sum Formulas

or
$$\sin a \sin b = \frac{1}{2}\left[\cos(a-b) - \cos(a+b)\right].$$

Likewise, using
$$\sin(a+b) = \sin a \cos b + \sin b \cos a$$

and
$$\sin(a-b) = \sin a \cos b - \sin b \cos a$$

, we obtain
$$\sin a \cos b = \frac{1}{2}\left[\sin(a+b) + \sin(a-b)\right]$$

and
$$\sin b \cos a = \frac{1}{2}\left[\sin(a+b) - \sin(a-b)\right].$$

□

Example 24

Find $\sin 45° \cos 75°$ and $\cos 45° \cos 75°$.

Solution.

We have
$$\begin{aligned}
\sin 45° \cos 75° &= \frac{1}{2}\left[\sin(45° + 75°) + \sin(45° - 75°)\right] \\
&= \frac{1}{2}\left[\sin 120° + \sin(-30°)\right] \\
&= \frac{1}{2}\left(\sin 60° - \sin 30°\right) \\
&= \frac{1}{2}\left(\frac{\sqrt{3}}{2} - \frac{1}{2}\right) \\
&= \frac{\sqrt{3}-1}{4}.
\end{aligned}$$

Additionally,
$$\begin{aligned}
\cos 45° \cos 75° &= \frac{1}{2}\left[\cos(45° - 75°) + \cos(45° + 75°)\right] \\
&= \frac{1}{2}\left[\cos(-30°) + \cos 120°\right]
\end{aligned}$$

$$= \frac{1}{2}\left(\cos 30° - \cos 60°\right)$$
$$= \frac{1}{2}\left(\frac{\sqrt{3}}{2} - \frac{1}{2}\right)$$
$$= \frac{\sqrt{3}-1}{4}.$$

5.6 Sum to Product Formulas

> **General 11**
>
> - $\cos p + \cos q = 2\cos\dfrac{p+q}{2}\cos\dfrac{p-q}{2}$
> - $\cos p - \cos q = -2\sin\dfrac{p+q}{2}\sin\dfrac{p-q}{2}$
> - $\sin p + \sin q = 2\sin\dfrac{p+q}{2}\cos\dfrac{p-q}{2}$
> - $\sin p - \sin q = 2\sin\dfrac{p-q}{2}\cos\dfrac{p+q}{2}$
> - $\tan p + \tan q = \dfrac{\sin(p+q)}{\cos p \cos q}$
> - $\tan p - \tan q = \dfrac{\sin(p-q)}{\cos p \cos q}.$

Proof. We have
$$\cos a \cos b = \frac{1}{2}\left[\cos(a-b) + \cos(a+b)\right]$$
or
$$\cos(a-b) + \cos(a+b) = 2\cos a \cos b.$$

Let $p = a+b$ and $q = a-b$. Then $a = \dfrac{p+q}{2}$ and $b = \dfrac{p-q}{2}$. It implies that $\cos p + \cos q = 2\cos\dfrac{p+q}{2}\cos\dfrac{p-q}{2}$. Similarly, we obtain
$$\cos p - \cos q = -2\sin\dfrac{p+q}{2}\sin\dfrac{p-q}{2}$$

5.6. Sum to Product Formulas

$$\sin p + \sin q = 2 \sin \frac{p+q}{2} \cos \frac{p-q}{2}$$

and

$$\sin p - \sin q = 2 \sin \frac{p-q}{2} \cos \frac{p+q}{2}.$$

In addition,

$$\tan p + \tan q = \frac{\sin p}{\cos p} + \frac{\sin q}{\cos q}$$
$$= \frac{\sin p \cos q + \sin q \cos p}{\cos p \cos q}$$
$$= \frac{\sin(p+q)}{\cos p \cos q}$$

and

$$\tan p - \tan q = \tan p + \tan(-q)$$
$$= \frac{\sin(p-q)}{\cos p \cos(-q)}$$
$$= \frac{\sin(p-q)}{\cos p \cos q}.$$

□

> **Example 25**
>
> Simplify the following expressions:
>
> 1. $\dfrac{\sin x + \sin 3x + \sin 5x}{\cos x + \cos 3x + \cos 5x}$;
>
> 2. $\dfrac{\sin x + \sin 3x + \sin 5x + \sin 7x}{\cos x + \cos 3x + \cos 5x + \cos 7x}$.

Solution. Simplify the following expressions:

1. $\dfrac{\sin x + \sin 3x + \sin 5x}{\cos x + \cos 3x + \cos 5x}$
   We have
   $$\frac{\sin x + \sin 3x + \sin 5x}{\cos x + \cos 3x + \cos 5x} = \frac{\sin x + \sin 5x + \sin 3x}{\cos x + \cos 5x + \cos 3x}$$
   $$= \frac{2\sin\left(\dfrac{x+5x}{2}\right)\cos\left(\dfrac{x-5x}{2}\right) + \sin 3x}{2\cos\left(\dfrac{x+5x}{2}\right)\cos\left(\dfrac{x-5x}{2}\right) + \cos 3x}$$

$$= \frac{2\sin 3x \cos 2x + \sin 3x}{2\cos 3x \cos 2x + \cos 3x}$$

$$= \frac{\sin 3x \left(2\cos 2x + 1\right)}{\cos 3x \left(2\cos 2x + 1\right)}$$

$$= \frac{\sin 3x}{\cos 3x}$$

$$= \tan 3x.$$

Therefore, $\dfrac{\sin x + \sin 3x + \sin 5x}{\cos x + \cos 3x + \cos 5x} = \tan 3x.$

2. $\dfrac{\sin x + \sin 3x + \sin 5x + \sin 7x}{\cos x + \cos 3x + \cos 5x + \cos 7x}$

We have

$$\frac{\sin x + \sin 3x + \sin 5x + \sin 7x}{\cos x + \cos 3x + \cos 5x + \cos 7x}$$

$$= \frac{\sin x + \sin 7x + \sin 3x + \sin 5x}{\cos x + \cos 7x + \cos 3x + \cos 5x}$$

$$= \frac{2\sin\left(\frac{x+7x}{2}\right)\cos\left(\frac{x-7x}{2}\right) + 2\sin\left(\frac{3x+5x}{2}\right)\cos\left(\frac{3x-5x}{2}\right)}{2\cos\left(\frac{x+7x}{2}\right)\cos\left(\frac{x-7x}{2}\right) + 2\cos\left(\frac{3x+5x}{2}\right)\cos\left(\frac{3x-5x}{2}\right)}$$

$$= \frac{2\sin 4x \cos 3x + 2\sin 4x \cos x}{2\cos 4x \cos 3x + 2\cos 4x \cos x}$$

$$= \frac{2\sin 4x \left(\cos 3x + \cos x\right)}{2\cos 4x \left(\cos 3x + \cos x\right)}$$

$$= \frac{\sin 4x}{\cos 4x}$$

$$= \tan 4x.$$

Therefore, $\dfrac{\sin x + \sin 3x + \sin 5x + \sin 7x}{\cos x + \cos 3x + \cos 5x + \cos 7x} = \tan 4x.$

> **Practice 1**
>
> Show that
>
> $$\frac{\sin x + \sin 3x + \sin 5x + \ldots + \sin(2n-1)x}{\cos x + \cos 3x + \cos 5x + \ldots + \cos(2n-1)x} = \tan nx.$$

Chapter 6

Trigonometric Equations

6.1 Sines

Let $\left(\overrightarrow{OP_0}, \overrightarrow{OP_1}\right) = \theta$ and $\left(\overrightarrow{OP_0}, \overrightarrow{OP_2}\right) = \pi - \theta$ (See Figure 6.1). It is not difficult to see that OP_1 and OP_2 are symmetric with respect to the vertical axis. Then θ and $\pi - \theta$ have the same value of sines. Namely, $\cos x = \cos \theta$ if and only if $\begin{bmatrix} x = \theta + 2k\pi \\ x = \pi - \theta + 2k\pi \end{bmatrix}$, where $k \in \mathbb{Z}$.

> **Example 26**
>
> Solve the following equations:
>
> 1. $\sin x = \dfrac{\sqrt{2}}{2}$.
>
>   We have $\sin x = \dfrac{\sqrt{2}}{2} = \sin \dfrac{\pi}{4}$.
>
>   It follows that $\begin{bmatrix} x = \dfrac{\pi}{4} + 2k\pi \\ x = \pi - \dfrac{\pi}{4} + 2k\pi \end{bmatrix}$.
>
>   Consequently, $\begin{bmatrix} x = \dfrac{\pi}{4} + 2k\pi \\ x = \dfrac{3\pi}{4} + 2k\pi \end{bmatrix}$, where $k \in \mathbb{Z}$.

Chapter 6. Trigonometric Equations

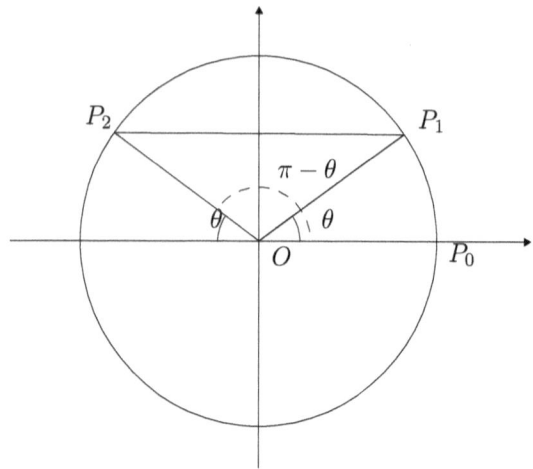

Figure 6.1: Angles θ and $\pi - \theta$

2. $\sin x \cos \dfrac{\pi}{3} - \sin \dfrac{\pi}{3} \cos x = \dfrac{1}{2}$.
   We have $\sin(a - b) = \sin a \cos b - \sin b \cos a$.
   The given equation is equivalent to
   $$\sin\left(x - \dfrac{\pi}{3}\right) = \sin \dfrac{\pi}{6}.$$
   Thus, $\left[\begin{array}{l} x - \dfrac{\pi}{3} = \dfrac{\pi}{6} + 2k\pi \\ x - \dfrac{\pi}{3} = \pi - \dfrac{\pi}{6} + 2k\pi \end{array}\right.$.

   Therefore, $\left[\begin{array}{l} x = \dfrac{\pi}{2} + 2k\pi \\ x = \dfrac{7\pi}{6} + 2k\pi \end{array}\right.$, where $k \in \mathbb{Z}$.

3. $\sin 2x = \sin\left(\dfrac{\pi}{3} - x\right)$.
   Since $\sin 2x = \sin\left(\dfrac{\pi}{3} - x\right)$, it follows that

6.2. Cosines

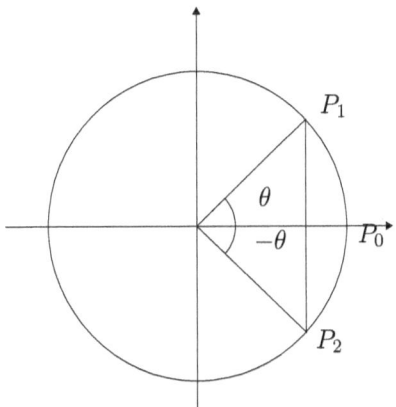

Figure 6.2:

$$\left[\begin{array}{l} 2x = \dfrac{\pi}{3} - x + 2k\pi \\ 2x = \pi - \left(\dfrac{\pi}{3} - x\right) + 2k\pi \end{array}\right. \text{ or } \left[\begin{array}{l} 3x = \dfrac{\pi}{3} + 2k\pi \\ x = \dfrac{2\pi}{3} + 2k\pi \end{array}\right..$$

Therefore, $\left[\begin{array}{l} x = \dfrac{\pi}{9} + \dfrac{2k\pi}{3} \\ x = \dfrac{2\pi}{3} + \dfrac{2k\pi}{3} \end{array}\right.$, where $k \in \mathbb{Z}$.

6.2 Cosines

In the unit circle, Let $\left(\overrightarrow{OP_0}, \overrightarrow{OP_1}\right) = \theta$ and $\left(\overrightarrow{OP_0}, \overrightarrow{OP_2}\right) = -\theta$.
It follows that OP_1 and OP_2 are symmetric with respect to the horizontal axis.
Consequently, θ and $-\theta$ have the same value of cosines.
Namely, $\cos x = \cos \theta$ if and only if $\left[\begin{array}{l} x = \theta + 2k\pi \\ x = -\theta + 2k\pi \end{array}\right.$, where $k \in \mathbb{Z}$.

Chapter 6. Trigonometric Equations

> **Example 27**
>
> Solve the following equations:
>
> 1. $\cos x = \dfrac{1}{2}$.
>
>   We have $\cos x = \dfrac{1}{2} = \cos \dfrac{\pi}{3}$.
>
>   Thus, $\left[\begin{array}{l} x = \dfrac{\pi}{3} + 2k\pi \\ x = -\dfrac{\pi}{3} + 2k\pi \end{array}\right.$, where $k \in \mathbb{Z}$.
>
> 2. $\cos x \cos \dfrac{\pi}{4} - \sin x \sin \dfrac{\pi}{4} = 1$
>
>   We have
> $$\cos x \cos \dfrac{\pi}{4} - \sin x \sin \dfrac{\pi}{4} = 1$$
>   or
> $$\cos\left(x + \dfrac{\pi}{4}\right) = 1.$$
>   Hence, $x + \dfrac{\pi}{4} = 2k\pi$.
>   Therefore, $x = -\dfrac{\pi}{4} + 2k\pi$, where $k \in \mathbb{Z}$.
>
> 3. $\sin^4 x - \cos^4 x = \dfrac{1}{2}$.
>
>   We have
> $$\begin{aligned}\sin^4 x - \cos^4 x &= \left(\sin^2 x - \cos^2 x\right)\left(\sin^2 x + \cos^2 x\right) \\ &= \left(\sin^2 x - \cos^2 x\right)(1) \\ &= -\left(\cos^2 x - \sin^2 x\right) \\ &= -\cos 2x.\end{aligned}$$
>
>   The given equation is equivalent to
> $$-\cos 2x = \dfrac{1}{2}$$
>   or
> $$\cos 2x = -\dfrac{1}{2} = \cos \dfrac{2\pi}{3}.$$

6.3. tangent

Then $\begin{bmatrix} 2x = \dfrac{2\pi}{3} + 2k\pi \\ 2x = -\dfrac{2\pi}{3} + 2k\pi \end{bmatrix}$.

Therefore, $\begin{bmatrix} x = \dfrac{\pi}{3} + k\pi \\ x = -\dfrac{\pi}{3} + k\pi \end{bmatrix}$, where $k \in \mathbb{Z}$.

4. $1 + 3\cos x + \cos 2x = \cos 3x + 2\sin x \sin 2x$.
Observe that

$$2\sin x \sin 2x = \cos(x - 2x) - \cos(x + 2x)$$
$$= \cos(-x) - \cos 3x = \cos x - \cos 3x$$

and
$$\cos 2x = 2\cos^2 x - 1.$$

The given equation can be written as

$$1 + 3\cos x + 2\cos^2 x - 1 = \cos 3x + \cos x - \cos 3x.$$

Then $2\cos^2 x + 2\cos x = 0$ or $2\cos x (\cos x + 1) = 0$.
It follows that $\begin{bmatrix} \cos x = 0 \\ \cos x + 1 = 0 \end{bmatrix}$ or $\begin{bmatrix} \cos x = 0 \\ \cos x = -1 \end{bmatrix}$.

Consequently, $\begin{bmatrix} x = \dfrac{\pi}{2} + 2k\pi \\ x = \pi + 2k\pi \end{bmatrix}$, where $k \in \mathbb{Z}$.

6.3 tangent

Let $\left(\overrightarrow{OP_0}, \overrightarrow{OP_1}\right) = \theta$. Basically, θ and $\pi + \theta$ have the same value of tangent. Consequently, $\tan x = \tan \theta$ if and only if $\begin{bmatrix} x = \theta + 2k\pi \\ x = \pi + \theta + 2k\pi \end{bmatrix}$, where $k \in \mathbb{Z}$. Therefore, the solutions of the equation can be written as $x = \theta + k\pi$, where $k \in \mathbb{Z}$.

> **Example 28**
>
> Solve the following equations:
>
> 1. $\tan x = -1$.
>   We have $\tan x = -1 = \tan\left(-\dfrac{\pi}{4}\right)$.
>   Consequently, $x = -\dfrac{\pi}{4} + k\pi$, where $k \in \mathbb{Z}$.
>
> 2. $\tan 3x = \tan\left(\dfrac{\pi}{3} - 2x\right)$.
>   We have $\tan 3x = \tan\left(\dfrac{\pi}{3} - 2x\right)$.
>   It follows that $3x = \dfrac{\pi}{3} - 2x + k\pi$ or $5x = \dfrac{\pi}{3} + k\pi$.
>   Hence, $x = \dfrac{\pi}{15} + \dfrac{k\pi}{5}$, where $k \in \mathbb{Z}$.
>
> 3. $\dfrac{\tan\dfrac{\pi}{4} - \tan x}{1 + \tan\dfrac{\pi}{4}\tan x} = \sqrt{3}$.
>
>   Since $\tan(a - b) = \dfrac{\tan a - \tan b}{1 + \tan a \tan b}$, the given equation is equivalent to
> $$\tan\left(\dfrac{\pi}{4} - x\right) = \sqrt{3} = \tan\dfrac{\pi}{3}.$$
>
>   Then $\dfrac{\pi}{4} - x = \dfrac{\pi}{3} + k\pi$.
>   Consequently, $x = \dfrac{\pi}{4} - \dfrac{\pi}{3} - k\pi = -\dfrac{\pi}{12} - k\pi$, where $k \in \mathbb{Z}$.

Chapter 7

Problems

Problem 1. Given that $\sin\alpha = \dfrac{1}{4}$ and $\sin\beta = \dfrac{1}{3}$ for $0 < \alpha, \beta < \dfrac{\pi}{2}$. Find the values of

1. $\sin(\alpha+\beta)$;
2. $\cos(\alpha+\beta)$;
3. $\tan(\alpha+\beta)$;
4. $\cot(\alpha+\beta)$.

Problem 2. Suppose that $\cos\theta = -\dfrac{2}{3}$, where $\dfrac{\pi}{2} < \theta < \pi$. Find the values of $\cos 2\theta$, $\sin\dfrac{\theta}{2}$, $\sin 3\theta$ and $\sin 4\theta$.

Problem 3. Given that $2\cos 2\theta - \cos\theta + 2 = 0$, where $-\dfrac{\pi}{2} < \theta < \dfrac{\pi}{2}$. Find the value of

1. $\tan\dfrac{\theta}{2}$;
2. $\cos 2\theta$;
3. $\sin 2\theta$.

Problem 4. Let $\theta = 36°$. Using the fact that $2\theta = 180° - 3\theta$, find the value of

Chapter 7. Problems

1. $\cos 36°$;
2. $\sin 36°$;
3. $\tan 36°$.

Problem 5. Suppose that $\sin\alpha + \cos\beta = \dfrac{1}{2}$ and $\cos\alpha + \sin\beta = \dfrac{1}{4}$, where $0 \leq \alpha, \beta \leq \dfrac{\pi}{2}$. Compute $\sin(\alpha+\beta)$ and $\tan(\alpha+\beta)$.

Problem 6. Suppose that $\sin\theta + \cos\theta = -\dfrac{1}{3}$, where $0 < \theta < \pi$. Find the values of

1. $\sin\theta\cos\theta$;
2. $\tan\theta + \dfrac{1}{\tan\theta}$;
3. $\sin^3\theta - \cos^3\theta$;
4. $\sin^3\theta + \cos^3\theta$;
5. $\sin^4\theta + \cos^4\theta$;
6. $\sin^4\theta - \cos^4\theta$.

Problem 7. Given that $\tan\theta = 3$. Find the values of the following expressions:

1. $A = \dfrac{\sin\theta + 2\cos\theta}{\sin\theta - 2\cos\theta}$;
2. $B = \dfrac{\sin^2\theta + 4\cos^2\theta}{\sin^2\theta - 3\cos^2\theta}$;
3. $C = \dfrac{\sin^3\theta - \cos^3\theta + \cos\theta}{\sin^3\theta + 2\cos^3\theta - \sin\theta}$;
4. $D = \dfrac{\sin^4\theta - 2\sin^2\theta - 7}{\cos^4\theta + \cos^2\theta - 3}$.

Problem 8. Simplify the following expressions:

1. $\sqrt{\tan^2\theta + \cot^2\theta + 2}$;
2. $\dfrac{\sin^2\theta - \tan^2\theta}{\cos^2\theta - \cot^2\theta}$;

3. $\sqrt{\sin^2\theta(1+\cot\theta)+\cos^2\theta(1+\tan\theta)}$;

4. $\dfrac{\cos^2\theta+\cos^2\theta\cot^2\theta}{\sin^2\theta+\sin^2\theta\tan^2\theta}$;

5. $\dfrac{1+\cos 2\theta}{\sin 2\theta}$;

6. $\dfrac{2\sin\theta\sin 2\theta+\cos 3\theta+3\cos\theta}{\sin\theta}$;

7. $\dfrac{2\cos 2\theta}{1+\cos 2\theta-\sin 2\theta}$.

Problem 9. Prove the following equalities:

1. $\dfrac{\sin\theta+\cos\theta-1}{\sin\theta-\cos\theta+1}=\dfrac{\cos\theta}{1+\sin\theta}$;

2. $\dfrac{\cos\theta\cot\theta-\sin\theta\tan\theta}{\dfrac{1}{\sin\theta}-\dfrac{1}{\cos\theta}}=1+\sin\theta\cos\theta$;

3. $(\tan\theta-\sin\theta)^2+(1-\cos\theta)^2=\left(\dfrac{1}{\cos\theta}-1\right)^2$;

4. $\dfrac{2\sin\theta\cos\theta-\cos\theta}{1-\sin\theta+\sin^2\theta-\cos^2\theta}=\dfrac{1}{\tan\theta}$;

5. $\left(\dfrac{\sin^4\theta-\tan^4\theta}{\cos^4\theta-\cot^4\theta}\right)\left(\dfrac{\sin^2\theta+1}{\cos^2\theta+1}\right)=\tan^{10}\theta$;

6. $\dfrac{\cos 2\theta}{\sqrt{2}\cos\left(\theta+\dfrac{\pi}{4}\right)}=\cos\theta-\sin\theta$.

Problem 10. Prove the following equalities:

1. $\dfrac{\sin^4\theta+2\sin\theta\cos\theta-\cos^4\theta}{\tan 2\theta-1}=\cos 2\theta$;

2. $\dfrac{\cos^3\theta-\cos 3\theta}{\cos\theta}+\dfrac{\sin^3\theta+\sin 3\theta}{\sin\theta}=3$;

3. $\dfrac{\sin 3\theta}{\sin^3\theta}+\dfrac{\cos 3\theta}{\cos^3\theta}=\dfrac{12\cot 2\theta}{\sin 2\theta}$.

Problem 11. Prove the following equalities:

1. $\sin 3\theta = 4 \sin \theta \sin \left(\dfrac{\pi}{3} + \theta\right) \sin \left(\dfrac{\pi}{3} - \theta\right)$;

2. $\cos 3\theta = 4 \cos \theta \cos \left(\dfrac{\pi}{3} + \theta\right) \cos \left(\dfrac{\pi}{3} - \theta\right)$;

3. $4 \sin \left(\theta + \dfrac{\pi}{3}\right) \sin \left(\theta - \dfrac{\pi}{3}\right) = 4\sin^2\theta - 3$;

4. $\sin(\theta - \gamma) \sin(\theta + \gamma) \cos(2\theta - 2\gamma) + \cos(4\theta - 2\gamma) + \cos 2\gamma$
$\qquad\qquad = \cos(2\theta - 4\gamma) + \cos 2\theta.$

Problem 12. For all integers k, prove that

1. $\sin(k\pi - \alpha) = (-1)^{k+1} \sin \alpha$;

2. $\cos(k\pi - \alpha) = (-1)^k \cos \alpha$;

3. $\tan(k\pi - \alpha) = -\tan \alpha$;

4. $\cot(k\pi - \alpha) = -\cot \alpha$.

Problem 13. Let A, B and C be the three angles of a triangle. For all integers k, prove the following identities:

1. $\sin kA = (-1)^{k+1} \sin k(B + C)$;

2. $\cos kA = (-1)^k \cos k(B + C)$;

3. $\tan kA = -\tan k(B + C)$;

4. $\cot kA = -\cot k(B + C)$.

Problem 14. Let ABC be a triangle. Prove the following identities:

1. $\sin(2k+1)\dfrac{A}{2} = (-1)^k \cos(2k+1)\left(\dfrac{B+C}{2}\right)$;

2. $\cos(2k+1)\dfrac{A}{2} = (-1)^k \sin(2k+1)\left(\dfrac{B+C}{2}\right)$;

3. $\tan(2k+1)\dfrac{A}{2} = \cot(2k+1)\dfrac{B}{2}$;

4. $\cot(2k+1)\dfrac{A}{2} = \tan(2k+1)\left(\dfrac{B+C}{2}\right)$.

Problem 15. Let ABC be a triangle. Prove the following equalities:

1. $\sin A + \sin B + \sin C = 4 \cos \dfrac{A}{2} \cos \dfrac{B}{2} \cos \dfrac{C}{2}$;

2. $\sin 2A + \sin 2B + \sin 2C = 4 \sin A \sin B \sin C$;

3. $\sin 3A + \sin 3B + \sin 3C = -4 \cos \dfrac{3A}{2} \cos \dfrac{3B}{2} \cos \dfrac{3C}{2}$;

4. $\sin 4A + \sin 4B + \sin 4C = -4 \sin 2A \sin 2B \sin 2C$;

5. $\sin (2k+1) A + \sin (2k+1) B + \sin (2k+1) C$;

$= (-1)^k \cos (2k+1) \dfrac{A}{2} \cos (2k+1) \dfrac{B}{2} \cos (2k+1) \dfrac{C}{2}$

6. $\sin 2kA + \sin 2kB + \sin 2kC = 4(-1)^{k+1} \sin kA \sin kB \sin kC$.

Problem 16. Let ABC be a triangle. Prove the following equalities:

1. $\cos A + \cos B + \cos C = 1 + 4 \sin \dfrac{A}{2} \sin \dfrac{B}{2} \sin \dfrac{C}{2}$;

2. $\cos 2A + \cos 2B + \cos 2C = -1 - 4 \cos A \cos B \cos C$;

3. $\cos 3A + \cos 3B + \cos 3C = -1 - 4 \sin \dfrac{3A}{2} \sin \dfrac{3B}{2} \sin \dfrac{3C}{2}$;

4. $\cos 4A + \cos 4B + \cos 4C = -1 + 4 \cos 2A \cos 2B \cos 2C$;

5. $\cos (2k+1) A + \cos (2k+1) B + \cos (2k+1) C$

$= 1 + 4(-1)^k \sin (2k+1) \dfrac{A}{2} \sin (2k+1) \dfrac{B}{2} \sin (2k+1) \dfrac{C}{2}$;

6. $\cos 2kA + \cos 2kB + \cos 2kC = -1 + 4(-1)^k \cos kA \cos kB \cos kC$.

Problem 17. Let ABC be a triangle. Prove the following equalities:

1. $\tan A + \tan B + \tan C = \tan A \tan B \tan C$;

2. $\tan 2A + \tan 2B + \tan 2C = \tan 2A \tan 2B \tan 2C$;

3. $\tan kA + \tan kB + \tan kC = \tan kA \tan kB \tan kC$;

4. $\tan\dfrac{A}{2}\tan\dfrac{B}{2}+\tan\dfrac{B}{2}\tan\dfrac{C}{2}+\tan\dfrac{C}{2}\tan\dfrac{A}{2}=1;$

5. $\tan(2k+1)\dfrac{A}{2}\tan(2k+1)\dfrac{B}{2}+\tan(2k+1)\dfrac{B}{2}\tan(2k+1)\dfrac{C}{2}$

$+\tan(2k+1)\dfrac{C}{2}\tan(2k+1)\dfrac{A}{2}=1.$

Problem 18. Let ABC be a triangle. Prove the following equalities:

1. $\cot A\cot B+\cot B\cot C+\cot C\cot A=1;$
2. $\cot kA\cot kB+\cot kB\cot kC+\cot kC\cot kA=1;$
3. $\cot\dfrac{A}{2}+\cot\dfrac{B}{2}+\cot\dfrac{C}{2}=\cot\dfrac{A}{2}\cot\dfrac{B}{2}\cot\dfrac{C}{2};$
4. $\cot(2k+1)\dfrac{A}{2}+\cot(2k+1)\dfrac{B}{2}+\cot(2k+1)\dfrac{C}{2}$

$=\cot(2k+1)\dfrac{A}{2}\cot(2k+1)\dfrac{B}{2}\cot(2k+1)\dfrac{C}{2}.$

Problem 19. Let ABC be a triangle. Show that

1. $\sin^3 A\cos(B-C)+\sin^3 B\cos(C-A)+\sin^3 C\cos(A-B)=3\sin A\sin B\sin C;$

2. $\sin^3 A\sin(B-C)+\sin^3 B\sin(C-A)+\sin^3 C\sin(A-B)=0;$

3. $\sin 3A\sin^3(B-C)+\sin 3B\sin^3(C-A)+\sin 3C\sin^3(A-B)=0;$

4. $\sin 3A\cos^3(B-C)+\sin 3B\cos^3(C-A)+\sin 3C\cos^3(A-B)=\sin 3A\sin 3B\sin 3C.$

Problem 20. Let x,y and z be three real numbers such that

$$x+y+z=0.$$

Prove the following equalities:

1. $\sin kx+\sin ky+\sin kz=-4\sin\dfrac{kx}{2}\sin\dfrac{ky}{2}\sin\dfrac{kz}{2};$

2. $\cos kx + \cos ky + \cos kz = 4\cos\dfrac{kx}{2}\cos\dfrac{ky}{2}\cos\dfrac{kz}{2} - 1$;

3. $\tan kx + \tan ky + \tan kz = \tan kx \tan ky \tan kz$;

4. $\cot kx \cot ky + \cot ky \cot kz + \cot kz \cot kx = 1$.

Problem 21. Let ABC be a triangle. Prove that

$$\sin^2\dfrac{A}{2}\cos(B-C) + \sin^2\dfrac{B}{2}\cos(C-A) + \sin^2\dfrac{C}{2}\cos(A-B)$$
$$= 2\cos\dfrac{A-B}{2}\cos\dfrac{B-C}{2}\cos\dfrac{C-A}{2} - 2\cos A \cos B \cos C - 1.$$

Problem 22. Let A, B and C be the three angles of a triangle. Prove that:

1. $\sin A + \sin B + \sin C \le \dfrac{3\sqrt{3}}{2}$;

2. $1 < \cos A + \cos B + \cos C \le \dfrac{3}{2}$;

3. $\tan A + \tan B + \tan C \ge 3\sqrt{3}$ (A, B and C are acute angles);

4. $\cot A + \cot B + \cot C \ge \sqrt{3}$;

5. $\sin^2 A + \sin^2 B + \sin^2 C \le \dfrac{9}{4}$;

6. $\cos^2 A + \cos^2 B + \cos^2 C \ge \dfrac{3}{4}$;

7. $\tan^2 A + \tan^2 B + \tan^2 C \ge 9$ (A, B and C are acute angles);

8. $1 < \sin\dfrac{A}{2} + \sin\dfrac{B}{2} + \sin\dfrac{C}{2} \le \dfrac{3}{2}$;

9. $2 < \cos\dfrac{A}{2} + \cos\dfrac{B}{2} + \cos\dfrac{C}{2} \le \dfrac{3\sqrt{3}}{2}$;

10. $\tan\dfrac{A}{2} + \tan\dfrac{B}{2} + \tan\dfrac{C}{2} \ge \sqrt{3}$;

11. $\dfrac{3}{4} \le \sin^2\dfrac{A}{2} + \sin^2\dfrac{B}{2} + \sin^2\dfrac{C}{2} < 1$;

12. $2 < \cos^2\dfrac{A}{2} + \cos^2\dfrac{B}{2} + \cos^2\dfrac{C}{2} \le \dfrac{9}{4}$;

13. $\tan^2 \frac{A}{2} + \tan^2 \frac{B}{2} + \tan^2 \frac{C}{2} \geq 1$;

14. $\sin A \sin B \sin C \leq \frac{3\sqrt{3}}{8}$;

15. $\cos A \cos B \cos C \leq \frac{1}{8}$;

16. $\sin \frac{A}{2} \sin \frac{B}{2} \sin \frac{C}{2} \leq \frac{1}{8}$;

17. $\cos \frac{A}{2} \cos \frac{A}{2} \cos \frac{A}{2} \leq \frac{3\sqrt{3}}{8}$;

18. $\tan \frac{A}{2} \tan \frac{B}{2} \tan \frac{C}{2} \leq \frac{1}{3\sqrt{3}}$.

Problem 23. Prove the following inequalities:

1. $\sin^4 \theta + \cos^4 \theta \geq \frac{1}{2}$;

2. $\sin^6 \theta + \cos^6 \theta \geq \frac{1}{4}$;

3. $\sin^8 \theta + \cos^8 \theta \geq \frac{1}{8}$;

4. $\sin^{2n} \theta + \cos^{2n} \theta \geq \frac{1}{2^{n-1}}$ for all positive integers n.

Problem 24. Prove the following equalities:

1. $\tan^2 \frac{\pi}{12} + \tan^2 \frac{3\pi}{12} + \tan^2 \frac{5\pi}{12} = 15$;

2. $\sin^4 \frac{\pi}{16} + \sin^4 \frac{3\pi}{16} + \sin^4 \frac{5\pi}{16} + \sin^4 \frac{7\pi}{16} = \frac{3}{2}$.

Problem 25. Simplify $\sin 4\theta - 4 \sin 3\theta + 6 \sin 2\theta - 4 \sin \theta$.

Problem 26. Calculate the following expressions:

1. $\tan 9° - \tan 27° - \tan 63° + \tan 81°$;

2. $\frac{1}{2 \sin 10°} - 2 \sin 70°$;

3. $3 \sin 15° \cos 15° + \frac{\sin 60°}{\sin^2 15° - \cos^2 15°}$.

Problem 27. Simplify the following expressions:

1. $\dfrac{1}{2}\tan\dfrac{\theta}{2} + \dfrac{1}{2^2}\tan\dfrac{\theta}{2^2} + \ldots + \dfrac{1}{2^n}\tan\dfrac{\theta}{2^n}$;

2. $\dfrac{1}{4\cos^2\dfrac{\theta}{2}} + \dfrac{1}{4^2\cos^2\dfrac{\theta}{2^2}} + \ldots + \dfrac{1}{4^n\cos^2\dfrac{\theta}{2^n}}$;

3. $\sin^3\dfrac{\theta}{3} + 3\sin^3\dfrac{\theta}{3^2} + 3^2\sin^3\dfrac{\theta}{3^3} + \ldots + 3^{n-1}\sin^3\dfrac{\theta}{3^n}$.

Problem 28. Prove that

$$\cos\frac{\pi}{15}\cos\frac{2\pi}{15}\cos\frac{3\pi}{15}\cos\frac{4\pi}{15}\cos\frac{5\pi}{15}\cos\frac{6\pi}{15}\cos\frac{7\pi}{15} = \frac{1}{2^7}.$$

Problem 29. Simplify the following expressions:

1. $A = \cos\dfrac{\theta}{2}\cos\dfrac{\theta}{2^2}\cos\dfrac{\theta}{2^3}\ldots\cos\dfrac{\theta}{2^n}$;

2. $B = (2\cos\theta - 1)(2\cos 2\theta - 1)(2\cos 2^2\theta - 1)\ldots(2\cos 2^n\theta - 1)$;

3. $C = \left(1 + \dfrac{1}{\cos\theta}\right)\left(1 + \dfrac{1}{\cos 2\theta}\right)\left(1 + \dfrac{1}{\cos 2^2\theta}\right)\ldots\left(1 + \dfrac{1}{\cos 2^n\theta}\right)$.

Problem 30. Given a triangle ABC. Prove that $\sin\dfrac{A}{2} \le \dfrac{a}{b+c}$, $\sin\dfrac{B}{2} \le \dfrac{b}{c+a}$ and $\sin\dfrac{C}{2} \le \dfrac{a+b}{2}$.

Problem 31. Show that $\dfrac{\sin(a-b)}{\cos a \cos b} + \dfrac{\sin(b-c)}{\cos b \cos c} + \dfrac{\sin(c-a)}{\cos c \cos a} = 0$.

Problem 32. Prove that $\cos\dfrac{\pi}{7} + \cos\dfrac{3\pi}{7} + \cos\dfrac{5\pi}{7} = \dfrac{1}{2}$.

Problem 33. Simplify the following expression:
$$S = \dfrac{1}{\cos a \cos(a+x)} + \dfrac{1}{\cos(a+x)\cos(a+2x)} + \ldots$$
$$+ \dfrac{1}{\cos[a+(n-1)x]\cos(a+nx)}.$$

Problem 34. Prove that $\sin a + \sin(a+x) + \sin(a+2x) + \ldots +$
$$\sin[a+(n-1)x] = \dfrac{\sin\dfrac{nx}{2}\cos\left[a + \left(\dfrac{n-1}{2}\right)x\right]}{\sin\dfrac{x}{2}}.$$

Problem 35. Prove that $\cos a + \cos(a+x) + \ldots + \cos[a+(n-1)x]$

$$= \frac{\sin\dfrac{nx}{2}\cos\left[a+\left(\dfrac{n-1}{2}\right)x\right]}{\sin\dfrac{x}{2}}.$$

Problem 36. Show that $\cos\dfrac{2\pi}{9}$, $\cos\dfrac{4\pi}{9}$, $\cos\dfrac{6\pi}{9}$ and $\cos\dfrac{8\pi}{9}$ are the roots of

$$16x^4 + 8x^3 - 12x^2 - 4x + 1 = 0.$$

Evaluate

$$\cos\frac{2\pi}{9} + \cos\frac{4\pi}{9} + \cos\frac{6\pi}{9} + \cos\frac{8\pi}{9}$$

and

$$\cos\frac{2\pi}{9}\cos\frac{4\pi}{9}\cos\frac{6\pi}{9}\cos\frac{8\pi}{9}.$$

Problem 37. Solve the following equations:

1. $\sin 3x + \sin x = 0$;
2. $\sin 4x + \sin 2x = 0$;
3. $\cos 3x + \cos 2x = 0$;
4. $\cos 5x + \cos 3x = 0$;
5. $\tan 4x + \tan x = 0$.

Problem 38. Solve the following equations:

1. $\sin^2 x + 2\sin x - 3 = 0$;
2. $2\cos^2 x + \cos x - 1 = 0$;
3. $\tan^2 x - \tan x - 2 = 0$.

Problem 39. Solve the following equations:

1. $2\cos^2 x - 3\sqrt{2}\cos x + 2 = 0$;
2. $\sin^2 x - 3\sin x + 2 = 0$;
3. $\cos^2\dfrac{x}{2} - \cos\dfrac{x}{2} - 2 = 0$;
4. $\tan^3 x - 3\tan^2 x + 3\tan x - 1 = 0$;

5. $\dfrac{1}{\sin^2 x} = \cot x + 3$;

6. $4 - \cos 2x - 7\sin x = 0$;

7. $\tan^2 \dfrac{x}{2} - \left(1 - \sqrt{3}\right) \tan \dfrac{x}{2} - \sqrt{3} = 0$.

Problem 40. Solve the following equations:

1. $\sin x + \sqrt{3}\cos x = 1$;

2. $\sin x + \cos x = 1$;

3. $\cos 2x - \sin 2x = -1$;

4. $\sqrt{3}\sin x + \cos x = \sqrt{2}$;

5. $\cos x - \sqrt{3}\sin x = 3$.

Problem 41. Solve the following equations:

1. $\sin^2 x + \sqrt{3}\sin x \cos x + 2\cos^2 x = 2$;

2. $3\sin^2 x - 2\sin x \cos x + 5\cos^2 x = 3$.

Problem 42. Let a, b, c be real numbers different from -1 and 1 and satisfy
$$a + b + c = abc.$$

Show that $\dfrac{a}{1-a^2} + \dfrac{b}{1-b^2} + \dfrac{c}{1-c^2} = \dfrac{4abc}{(1-a^2)(1-b^2)(1-c^2)}$.

Problem 43. Given real numbers x, y and z such that x, y and $z > 0$. Prove that
$$\dfrac{x}{x + \sqrt{(x+y)(x+z)}} + \dfrac{y}{y + \sqrt{(y+z)(y+x)}} + \dfrac{z}{z + \sqrt{(z+x)(z+y)}} \leq 1.$$

Problem 44. Suppose that x, y and z are positive real numbers that satisfy $x + y + z = xyz$.
Prove that

1. $\dfrac{1}{\sqrt{1+x^2}} + \dfrac{1}{\sqrt{1+y^2}} + \dfrac{1}{\sqrt{1+z^2}} \leq \dfrac{3}{2}$;

2. $\dfrac{x}{\sqrt{1+x^2}} + \dfrac{y}{\sqrt{1+y^2}} + \dfrac{z}{\sqrt{1+z^2}} \leq \dfrac{3\sqrt{3}}{2}$.

Problem 45. Solve the following system of equations:
$$\begin{cases} x+y = \dfrac{\pi}{2} \\ \sin x + \sin y = 1 \end{cases}.$$

Problem 46. Solve the following system of equations:
$$\begin{cases} x+y = \dfrac{\pi}{2} \\ \tan x + \tan y = 2 \end{cases}.$$

Problem 47. Let x, y and z be real numbers such that $\sin x + \sin y + \sin z = 0$ and $\cos x + \cos y + \cos z = 0$. Prove that

1. $\cos(x-y) = -\dfrac{1}{2}$;

2. $\cos(\theta - x) + \cos(\theta - y) + \cos(\theta - z) = 0$ for all $\theta \in \mathbb{R}$;

3. $\sin^2 x + \sin^2 y + \sin^2 z = \dfrac{3}{2}$;

4. $2\left(\cot^2 x \cot^2 y + \cot^2 y \cot^2 z + \cot^2 z \cot^2 x\right) = 9\cot^2 x \cot^2 y \cot^2 z$.

Problem 48. Let (u_n) be a sequence such that $u_1 = \dfrac{\sqrt{2}}{2}$ and $u_{n+1} = \dfrac{\sqrt{2}}{2}\sqrt{1 - \sqrt{1 - u_n^2}}$. Find u_n in terms of n.

Problem 49. Suppose that $\tan x_1 \tan x_2 ... \tan x_n = k$. Find the maximum value of
$$A = \sin x_1 \sin x_2 ... \sin x_n.$$

Problem 50. Compute $S = \tan^6 \dfrac{\pi}{18} + \tan^6 \dfrac{5\pi}{18} + \tan^6 \dfrac{7\pi}{18}$.

Problem 51. Let (a_n) be a sequence defined by $a_n = \tan n° \tan(n-1)°$. Compute $S_n = \displaystyle\sum_{k=1}^{n} a_k$ in terms of n.

Problem 52. Let
$$S_n = \tan\theta \tan^2\dfrac{\theta}{2} + 2\tan\dfrac{\theta}{2}\tan^2\dfrac{\theta}{2^2} + ... + 2^{n-1}\tan\dfrac{\theta}{2^{n-1}}\tan^2\dfrac{\theta}{2^n}.$$

Prove that $S_n = \tan\theta - 2^n \tan\dfrac{\theta}{2^n}$.

Problem 53. Let $a_1, a_2, ..., a_n$ be real numbers such that $0 < a_1 < a_2 < ... < a_n < \frac{\pi}{2}$. Prove that

$$\tan a_1 < \frac{\sin a_1 + \sin a_2 + ... + \sin a_n}{\cos a_1 + \cos a_2 + ... + \cos a_n} < \tan a_n.$$

Problem 54. Simplify $S = \dfrac{1}{\sin 2\theta} + \dfrac{1}{\sin 2^2\theta} + \dfrac{1}{\sin 2^3\theta} + ... + \dfrac{1}{\sin 2^n\theta}$.

Problem 55. Find the product

$$P = \left(1 - \tan^2 \frac{\theta}{2}\right)\left(1 - \tan^2 \frac{\theta}{2^2}\right) ... \left(1 - \tan^2 \frac{\theta}{2^n}\right).$$

Problem 56. Simplify the following expressions:

1. $S_1 = 1 + \dfrac{\cos \theta}{\cos \theta} + \dfrac{\cos 2\theta}{\cos^2 \theta} + ... + \dfrac{\cos n\theta}{\cos^n \theta}$;

2. $S_2 = \dfrac{\sin \theta}{\cos \theta} + \dfrac{\sin 2\theta}{\cos^2 \theta} + ... + \dfrac{\sin n\theta}{\cos^n \theta}$.

Problem 57. Let r and R be the radii of the inscribed circle and the circumscribe circle of triangle ABC respectively. Prove that

$$\frac{1}{\sin \frac{A}{2}} + \frac{1}{\sin \frac{B}{2}} + \frac{1}{\sin \frac{C}{2}} \geq 4\sqrt{\frac{R}{r}}.$$

Chapter 7. Problems

Chapter 8

Solutions

Problem 1. Given that $\sin\alpha = \dfrac{1}{4}$ and $\sin\beta = \dfrac{1}{3}$ for $0 < \alpha, \beta < \dfrac{\pi}{2}$. Find the values of

1. $\sin(\alpha + \beta)$;

2. $\cos(\alpha + \beta)$;

3. $\tan(\alpha + \beta)$;

4. $\cot(\alpha + \beta)$.

Solution.   1. Find $\sin(\alpha + \beta)$.
Using the fact that $\sin^2\alpha + \cos^2\alpha = 1$ or $\cos^2\alpha = 1 - \sin^2\alpha$.
We obtain $\cos^2\alpha = 1 - \left(\dfrac{1}{4}\right)^2 = 1 - \dfrac{1}{16} = \dfrac{15}{16}$.
Since $0 < \alpha < \dfrac{\pi}{2}$, it follows that $\cos\alpha > 0$.
Then $\cos\alpha = \sqrt{\dfrac{15}{16}} = \dfrac{\sqrt{15}}{4}$.
Likewise, $\cos\beta = \dfrac{2\sqrt{2}}{3}$.
It implies that

$$\sin(\alpha + \beta) = \sin\alpha\cos\beta + \sin\beta\cos\alpha$$

$$= \left(\dfrac{1}{4}\right)\left(\dfrac{2\sqrt{2}}{3}\right) + \left(\dfrac{1}{3}\right)\left(\dfrac{\sqrt{15}}{4}\right)$$

$$= \frac{2\sqrt{2}}{12} + \frac{\sqrt{15}}{12}$$
$$= \frac{2\sqrt{2} + \sqrt{15}}{12}.$$

Therefore, $\sin(\alpha + \beta) = \dfrac{2\sqrt{2} + \sqrt{15}}{12}$.

2. Find $\cos(\alpha + \beta)$.
   We have
   $$\cos(\alpha + \beta) = \cos\alpha\cos\beta - \sin\alpha\sin\beta$$
   $$= \left(\frac{\sqrt{15}}{4}\right)\left(\frac{2\sqrt{2}}{3}\right) - \left(\frac{1}{4}\right)\left(\frac{1}{3}\right)$$
   $$= \frac{2\sqrt{30}}{12} - \frac{1}{12}$$
   $$= \frac{2\sqrt{30} - 1}{12}.$$

   Therefore, $\cos(\alpha + \beta) = \dfrac{2\sqrt{30} - 1}{12}$.

3. Find $\tan(\alpha + \beta)$.
   We have
   $$\tan\alpha = \frac{\sin\alpha}{\cos\alpha} = \frac{\frac{1}{4}}{\frac{\sqrt{15}}{4}} = \frac{1}{\sqrt{15}} = \frac{\sqrt{15}}{15}$$

   and
   $$\tan\beta = \frac{\sin\beta}{\cos\beta} = \frac{\frac{1}{3}}{\frac{2\sqrt{2}}{3}} = \frac{1}{2\sqrt{2}} = \frac{\sqrt{2}}{4}.$$

   It follows that
   $$\tan(\alpha + \beta) = \frac{\tan\alpha + \tan\beta}{1 - \tan\alpha\tan\beta}$$
   $$= \frac{\frac{\sqrt{15}}{15} + \frac{\sqrt{2}}{4}}{1 - \left(\frac{\sqrt{15}}{15}\right)\left(\frac{\sqrt{2}}{4}\right)}$$

$$= \frac{\frac{4\sqrt{15}+15\sqrt{2}}{60}}{1-\frac{\sqrt{30}}{60}}$$

$$= \frac{\frac{4\sqrt{15}+15\sqrt{2}}{60}}{\frac{60-\sqrt{30}}{60}}$$

$$= \frac{4\sqrt{15}+15\sqrt{2}}{60-\sqrt{30}}.$$

Therefore, $\tan(\alpha+\beta) = \dfrac{4\sqrt{15}+15\sqrt{2}}{60-\sqrt{30}}$.

4. Find $\cot(\alpha+\beta)$.

Since $\cot x$ is the reciprocal of $\tan x$,

$$\cot(\alpha+\beta) = \frac{1}{\tan(\alpha+\beta)}$$

$$= \frac{1}{\frac{4\sqrt{15}+15\sqrt{2}}{60-\sqrt{30}}}$$

$$= \frac{60-\sqrt{30}}{4\sqrt{15}+15\sqrt{2}}.$$

Therefore, $\cot(\alpha+\beta) = \dfrac{60-\sqrt{30}}{4\sqrt{15}+15\sqrt{2}}$.

Problem 2. Suppose that $\cos\theta = -\dfrac{2}{3}$, where $\dfrac{\pi}{2} < \theta < \pi$. Find the values of $\cos 2\theta$, $\sin\dfrac{\theta}{2}$, $\sin 3\theta$ and $\sin 4\theta$.

Solution. • Find $\cos 2\theta$.
We have $\cos 2\theta = 2\cos^2\theta - 1$.
By knowing that $\cos\theta = -\dfrac{2}{3}$, then

$$\cos 2\theta = 2\left(-\frac{2}{3}\right)^2 - 1 = 2\left(\frac{4}{9}\right) - 1 = \frac{8}{9} - 1 = -\frac{1}{9}.$$

Therefore, $\cos 2\theta = -\dfrac{1}{9}$.

Chapter 8. Solutions

- Find $\sin\dfrac{\theta}{2}$.
  We have
  $$\begin{aligned}\sin^2\dfrac{\theta}{2} &= \dfrac{1-\cos\theta}{2}\\ &= \dfrac{1-\left(-\dfrac{2}{3}\right)}{2}\\ &= \dfrac{1+\dfrac{2}{3}}{2}\\ &= \dfrac{\dfrac{5}{3}}{2}\\ &= \dfrac{5}{6}.\end{aligned}$$

  It follows that $\sin\dfrac{\theta}{2} = \pm\sqrt{\dfrac{5}{6}} = \pm\dfrac{\sqrt{30}}{6}$.
  Since $\dfrac{\pi}{2} < \theta < \pi$ or $\dfrac{\pi}{4} < \dfrac{\theta}{2} < \dfrac{\pi}{2}$, then $\sin\dfrac{\theta}{2} > 0$.
  Consequently, $\sin\dfrac{\theta}{2} = \dfrac{\sqrt{30}}{6}$.

- Find $\sin 3\theta$.
  Observe that $\sin 3\theta = 3\sin\theta - 4\sin^3\theta$.
  Moreover $\sin^2\theta + \cos^2\theta = 1$ or $\sin^2\theta = 1 - \cos^2\theta$.
  It implies that $\sin^2\theta = 1 - \left(-\dfrac{2}{3}\right)^2 = 1 - \dfrac{4}{9} = \dfrac{5}{9}$.
  Then $\sin\theta = \dfrac{\sqrt{5}}{3}$ because $\dfrac{\pi}{2} < \theta < \pi$. It turns out that
  $$\begin{aligned}\sin 3\theta &= 3\left(\dfrac{\sqrt{5}}{3}\right) - 4\left(\dfrac{\sqrt{5}}{3}\right)^3\\ &= \sqrt{5} - \dfrac{20\sqrt{5}}{27} = \dfrac{27\sqrt{5} - 20\sqrt{5}}{27}\\ &= \dfrac{7\sqrt{5}}{27}.\end{aligned}$$

  Therefore, $\sin 3\theta = \dfrac{7\sqrt{5}}{27}$.

- Find $\sin 4\theta$.

    We have $\sin 4\theta = 2\sin 2\theta \cos 2\theta$
    $$= 4\sin\theta \cos\theta \cos 2\theta$$
    $$= 4\left(\frac{\sqrt{5}}{3}\right)\left(-\frac{2}{3}\right)\left(-\frac{1}{9}\right)$$
    $$= -\frac{8\sqrt{5}}{81}.$$

    Therefore, $\sin 4\theta = -\dfrac{8\sqrt{5}}{81}$.

Problem 3. Given that $2\cos 2\theta - \cos\theta + 2 = 0$, where $-\dfrac{\pi}{2} < \theta < \dfrac{\pi}{2}$. Find the value of

1. $\tan\dfrac{\theta}{2}$;

2. $\cos 2\theta$;

3. $\sin 2\theta$.

Solution. Find the value of

1. $\tan\dfrac{\theta}{2}$

    Since $2\cos 2\theta - \cos\theta + 2 = 0$ and $\cos 2\theta = 2\cos^2\theta - 1$, it follows that
    $$2\left(2\cos^2\theta - 1\right) - \cos\theta + 2 = 0$$

    or
    $$4\cos^2\theta - \cos\theta = 0.$$

    Then $\cos\theta(4\cos\theta - 1) = 0$.

    Hence, $\begin{bmatrix} \cos\theta = 0 \\ 4\cos\theta - 1 = 0 \end{bmatrix}$ or $\begin{bmatrix} \cos\theta = 0 \\ \cos\theta = \dfrac{1}{4} \end{bmatrix}$.

    Moreover, $-\dfrac{\pi}{2} < \theta < \dfrac{\pi}{2}$. Then $\cos\theta > 0$.

    It implies that $\cos\theta = \dfrac{1}{4}$.

75

Using half-angle formula,

$$\tan^2\frac{\theta}{2} = \frac{1-\cos\theta}{1+\cos\theta} = \frac{1-\frac{1}{4}}{1+\frac{1}{4}} = \frac{\frac{3}{4}}{\frac{5}{4}} = \frac{3}{5}.$$

Thus, $\tan\frac{\theta}{2} = \pm\sqrt{\frac{3}{5}} = \pm\frac{\sqrt{15}}{5}.$

2. $\cos 2\theta$

Using double-angle formula, we have $\cos 2\theta = 2\cos^2\theta - 1$. It follows that

$$\begin{aligned}\cos 2\theta &= 2\left(\frac{1}{4}\right)^2 - 1 \\ &= 2\left(\frac{1}{16}\right) - 1 \\ &= \frac{1}{8} - 1 \\ &= \frac{1}{8} - \frac{8}{8} \\ &= \frac{1-8}{8} \\ &= -\frac{7}{8}.\end{aligned}$$

Therefore, $\cos 2\theta = -\frac{7}{8}.$

3. $\sin 2\theta$

Using the fact that $\sin^2\theta + \cos^2\theta = 1$, we obtain

$$\begin{aligned}\sin^2\theta &= 1 - \cos^2\theta \\ &= 1 - \left(\frac{1}{4}\right)^2 \\ &= 1 - \frac{1}{16} \\ &= \frac{16}{16} - \frac{1}{16} \\ &= \frac{16-1}{16}\end{aligned}$$

$$= \frac{15}{16}.$$

It implies that $\sin\theta = \pm\sqrt{\frac{15}{16}} = \pm\frac{\sqrt{15}}{4}$.
From double-angle formula, we obtain

$$\sin 2\theta = 2\sin\theta\cos\theta$$
$$= 2\left(\pm\frac{\sqrt{15}}{4}\right)\left(\frac{1}{4}\right)$$
$$= \pm\frac{\sqrt{15}}{8}.$$

Therefore, $\sin 2\theta = \pm\dfrac{\sqrt{15}}{8}$.

Problem 4. Let $\theta = 36°$. Using the fact that $2\theta = 180° - 3\theta$, find the value of

1. $\cos 36°$;

2. $\sin 36°$;

3. $\tan 36°$.

Solution. Find the value of

1. $\cos 36°$
   We have $2\theta = 180° - 3\theta$. Then

$$\cos 2\theta = \cos(180° - 3\theta) = -\cos 3\theta.$$

By knowing that $\cos 2\theta = 2\cos^2\theta - 1$ and $\cos 3\theta = 4\cos^3\theta - 3\cos\theta$, it follows that

$$2\cos^2\theta - 1 = -\left(4\cos^3\theta - 3\cos\theta\right)$$
$$4\cos^3\theta + 2\cos^2\theta - 3\cos\theta - 1 = 0$$
$$4\cos^3\theta + 4\cos^2\theta - 2\cos^2\theta - 2\cos\theta - \cos\theta - 1 = 0$$
$$4\cos^2\theta(\cos\theta + 1) - 2\cos\theta(\cos\theta + 1) - (\cos\theta + 1) = 0$$
$$(\cos\theta + 1)\left(4\cos^2\theta - 2\cos\theta - 1\right) = 0.$$

Since $0 < \cos 36° < 1$, then $4\cos^2\theta - 2\cos\theta - 1 = 0$. The discriminant of the last equation is
$$\Delta' = b'^2 - ac = (-1)^2 - (4)(-1) = 1 + 4 = 5.$$
We obtain $\cos\theta = \dfrac{-b' \pm \sqrt{\Delta'}}{a} = \dfrac{1 \pm \sqrt{5}}{4}$.

Therefore, $\cos\theta = \dfrac{1+\sqrt{5}}{4}$.

2. $\sin 36°$

Using the fact that $\sin^2 36° + \cos^2 36° = 1$, we obtain
$$\begin{aligned}\sin^2 36° &= 1 - \cos^2 36° \\ &= 1 - \left(\dfrac{1+\sqrt{5}}{4}\right)^2 \\ &= 1 - \dfrac{1 + 2\sqrt{5} + \sqrt{5}^2}{16} \\ &= \dfrac{16 - 1 - 2\sqrt{5} - 5}{16} \\ &= \dfrac{10 - 2\sqrt{5}}{16}.\end{aligned}$$

It follows that $\sin 36° = \sqrt{\dfrac{10 - 2\sqrt{5}}{16}} = \dfrac{\sqrt{10 - 2\sqrt{5}}}{4}$ since $\sin 36° > 0$. Therefore, $\sin 36° = \dfrac{\sqrt{10 - 2\sqrt{5}}}{4}$.

3. $\tan 36°$

We have
$$\begin{aligned}\tan 36° &= \dfrac{\sin 36°}{\cos 36°} \\ &= \dfrac{\frac{\sqrt{10 - 2\sqrt{5}}}{4}}{\frac{1+\sqrt{5}}{4}} \\ &= \dfrac{\sqrt{10 - 2\sqrt{5}}}{4} \times \dfrac{4}{1+\sqrt{5}} \\ &= \dfrac{\sqrt{10 - 2\sqrt{5}}}{\sqrt{5}+1}.\end{aligned}$$

Therefore, $\tan 36° = \dfrac{\sqrt{10-2\sqrt{5}}}{\sqrt{5}+1}$.

Problem 5. Suppose that $\sin\alpha+\cos\beta = \dfrac{1}{2}$ and $\cos\alpha+\sin\beta = \dfrac{1}{4}$, where $0 \le \alpha, \beta \le \dfrac{\pi}{2}$. Compute $\sin(\alpha+\beta)$ and $\tan(\alpha+\beta)$.

Solution. • Compute $\sin(\alpha+\beta)$.
We have $\sin\alpha + \cos\beta = \dfrac{1}{2}$. Then

$$(\sin\alpha + \cos\beta)^2 = \left(\dfrac{1}{2}\right)^2$$

or
$$\sin^2\alpha + 2\sin\alpha\cos\beta + \cos^2\beta = \dfrac{1}{4}. \qquad (1)$$

Moreover, $\cos\alpha + \sin\beta = \dfrac{1}{4}$. Then

$$(\cos\alpha + \sin\beta)^2 = \left(\dfrac{1}{4}\right)^2$$

or
$$\cos^2\alpha + 2\sin\beta\cos\alpha + \sin^2\beta = \dfrac{1}{16}. \qquad (2)$$

Adding (1) and (2), we obtain
$\sin^2\alpha + \cos^2\alpha + \sin^2\beta + \cos^2\beta + 2(\sin\alpha\cos\beta + \sin\beta\cos\alpha) = \dfrac{5}{16}$
$1 + 1 + 2(\sin\alpha\cos\beta + \sin\beta\cos\alpha) = \dfrac{5}{16}$
$2 + 2\sin(\alpha+\beta) = \dfrac{5}{16}$.
It follows that $2\sin(\alpha+\beta) = \dfrac{5}{16} - 2 = -\dfrac{27}{16}$.
Then $\sin(\alpha+\beta) = -\dfrac{27}{32}$.
Therefore, $\sin(\alpha+\beta) = -\dfrac{27}{32}$.

• Compute $\tan(\alpha+\beta)$.
Using the fact that $\sin^2(\alpha+\beta) + \cos^2(\alpha+\beta) = 1$, then

$$\cos^2(\alpha+\beta) = 1 - \sin^2(\alpha+\beta)$$

$$= 1 - \left(-\frac{27}{32}\right)^2$$
$$= \frac{32^2 - 27^2}{32^2}$$
$$= \frac{5 \times 69}{32^2}$$
$$= \frac{345}{32^2}.$$

It follows that $\cos(\alpha + \beta) = \pm\sqrt{\dfrac{345}{32^2}} = \pm\dfrac{\sqrt{345}}{32}$.
We obtain

$$\tan(\alpha + \beta) = \frac{\sin(\alpha + \beta)}{\cos(\alpha + \beta)}$$
$$= \frac{-\dfrac{27}{32}}{\pm\dfrac{\sqrt{345}}{32}}$$
$$= \mp\frac{24}{\sqrt{345}}$$
$$= \mp\frac{24\sqrt{345}}{345}.$$

Therefore, $\tan(\alpha + \beta) = \mp\dfrac{24\sqrt{345}}{345}$.

Problem 6. Suppose that $\sin\theta + \cos\theta = -\dfrac{1}{3}$, where $0 < \theta < \pi$. Find the values of

1. $\sin\theta\cos\theta$;

2. $\tan\theta + \dfrac{1}{\tan\theta}$;

3. $\sin^3\theta - \cos^3\theta$;

4. $\sin^3\theta + \cos^3\theta$;

5. $\sin^4\theta + \cos^4\theta$;

6. $\sin^4\theta - \cos^4\theta$.

Solution. Find the values of

1. $\sin\theta\cos\theta$

   We have $\sin\theta + \cos\theta = -\dfrac{1}{3}$. Then
   $$(\sin\theta + \cos\theta)^2 = \dfrac{1}{9}$$
   or
   $$\sin^2\theta + \cos^2\theta + 2\sin\theta\cos\theta = \dfrac{1}{9}.$$
   By knowing that $\sin^2\theta + \cos^2\theta = 1$, we obtain
   $$1 + 2\sin\theta\cos\theta = \dfrac{1}{9}.$$
   It follows that $2\sin\theta\cos\theta = \dfrac{1}{9} - 1 = -\dfrac{8}{9}.$
   Consequently, $\sin\theta\cos\theta = -\dfrac{4}{9}.$

2. $\tan\theta + \dfrac{1}{\tan\theta}$

   We have
   $$\begin{aligned}\tan\theta + \dfrac{1}{\tan\theta} &= \dfrac{\sin\theta}{\cos\theta} + \dfrac{\cos\theta}{\sin\theta}\\ &= \dfrac{\sin^2\theta + \cos^2\theta}{\sin\theta\cos\theta}\\ &= \dfrac{1}{\sin\theta\cos\theta}.\end{aligned}$$
   Since $\sin\theta\cos\theta = -\dfrac{4}{9}$, it implies that
   $$\tan\theta + \dfrac{1}{\tan\theta} = \dfrac{1}{-\dfrac{4}{9}} = -\dfrac{9}{4}.$$

3. $\sin^3\theta - \cos^3\theta$

   Observe that
   $$\begin{aligned}\sin^3\theta - \cos^3\theta &= (\sin\theta - \cos\theta)\left(\sin^2\theta + \sin\theta\cos\theta + \cos^2\theta\right)\\ &= (\sin\theta - \cos\theta)\left(1 - \dfrac{4}{9}\right)\end{aligned}$$

$$= \frac{5}{9}(\sin\theta - \cos\theta).$$

Moreover,
$$(\sin\theta - \cos\theta)^2 = \sin^2\theta - 2\sin\theta\cos\theta + \cos^2\theta$$
$$= 1 - 2\left(-\frac{5}{9}\right)$$
$$= 1 + \frac{10}{9} = \frac{19}{9}.$$

Then $\sin\theta - \cos\theta = \pm\frac{\sqrt{19}}{3}$.

However, $\sin\theta\cos\theta = -\frac{5}{9} < 0$ and $0 < \theta < \pi$. Then θ is in the second quadrant of the unit circle. Then $\sin\theta > 0$ and $\cos\theta < 0$. It follows that $\sin\theta - \cos\theta > 0$.

As a result, $\sin\theta - \cos\theta = \frac{\sqrt{19}}{3}$.

Consequently, $\sin^3\theta - \cos^3\theta = \frac{5}{9}\left(\frac{\sqrt{19}}{3}\right) = \frac{5\sqrt{19}}{27}$.

4. $\sin^3\theta + \cos^3\theta$;
We have
$$\sin^3\theta + \cos^3\theta = (\sin\theta + \cos\theta)\left(\sin^2\theta - \sin\theta\cos\theta + \cos^2\theta\right)$$
$$= (\sin\theta + \cos\theta)(1 - \sin\theta\cos\theta).$$

We know that $\sin\theta + \cos\theta = -\frac{1}{3}$ and $\sin\theta\cos\theta = -\frac{4}{9}$.
It follows that
$$\sin^3\theta + \cos^3\theta = \left(-\frac{1}{3}\right)\left(1 + \frac{4}{9}\right)$$
$$= \left(-\frac{1}{3}\right)\left(\frac{13}{9}\right)$$
$$= -\frac{13}{27}.$$

Therefore, $\sin^3\theta + \cos^3\theta = -\frac{13}{27}$.

5. $\sin^4\theta + \cos^4\theta$;
Observe that
$$\sin^4\theta + \cos^4\theta = \left(\sin^2\theta\right)^2 + \left(\cos^2\theta\right)^2$$

$$\begin{aligned}&= \left(\sin^2\theta\right)^2 + 2\sin^2\theta\cos^2\theta + \left(\cos^2\theta\right)^2 - 2\sin^2\theta\cos^2\theta\\&= \left(\sin^2\theta + \cos^2\theta\right)^2 - 2(\sin\theta\cos\theta)^2\\&= 1 - 2(\sin\theta\cos\theta)^2.\end{aligned}$$

We know that $\sin\theta\cos\theta = -\dfrac{4}{9}$.
It implies that

$$\begin{aligned}\sin^4\theta + \cos^4\theta &= 1 - 2\left(-\frac{4}{9}\right)^2\\&= 1 - 2\left(\frac{16}{81}\right)\\&= 1 - \frac{32}{81}\\&= \frac{49}{81}.\end{aligned}$$

Therefore, $\sin^4\theta + \cos^4\theta = \dfrac{49}{81}$.

6. $\sin^4\theta - \cos^4\theta$
We have

$$\begin{aligned}\sin^4\theta - \cos^4\theta &= \left(\sin^2\theta\right)^2 - \left(\cos^2\theta\right)^2\\&= \left(\sin^2\theta - \cos^2\theta\right)\left(\sin^2\theta + \cos^2\theta\right)\\&= (\sin\theta - \cos\theta)(\sin\theta + \cos\theta)(1)\\&= (\sin\theta - \cos\theta)(\sin\theta + \cos\theta).\end{aligned}$$

By knowing that $\sin\theta - \cos\theta = \dfrac{\sqrt{19}}{3}$ and $\sin\theta + \cos\theta = -\dfrac{1}{3}$, it implies that

$$\begin{aligned}\sin^4\theta - \cos^4\theta &= \left(\frac{\sqrt{19}}{3}\right)\left(-\frac{1}{3}\right)\\&= -\frac{\sqrt{19}}{9}.\end{aligned}$$

Therefore, $\sin^4\theta - \cos^4\theta = -\dfrac{\sqrt{19}}{9}$.

Problem 7. Given that $\tan\theta = 3$. Find the values of the following expressions:

1. $A = \dfrac{\sin\theta + 2\cos\theta}{\sin\theta - 2\cos\theta}$;

2. $B = \dfrac{\sin^2\theta + 4\cos^2\theta}{\sin^2\theta - 3\cos^2\theta}$;

3. $C = \dfrac{\sin^3\theta - \cos^3\theta + \cos\theta}{\sin^3\theta + 2\cos^3\theta - \sin\theta}$;

4. $D = \dfrac{\sin^4\theta - 2\sin^2\theta - 7}{\cos^4\theta + \cos^2\theta - 3}$.

Solution. Find the values of the following expressions:

1. $A = \dfrac{\sin\theta + 2\cos\theta}{\sin\theta - 2\cos\theta}$;

We have
$$A = \dfrac{\sin\theta + 2\cos\theta}{\sin\theta - 2\cos\theta}$$
$$= \dfrac{\dfrac{\sin\theta + 2\cos\theta}{\sin\theta}}{\dfrac{\sin\theta - 2\cos\theta}{\cos\theta}}$$
$$= \dfrac{\dfrac{\sin\theta}{\sin\theta} + 2 \times \dfrac{\cos\theta}{\sin\theta}}{\dfrac{\sin\theta}{\cos\theta} - 2 \times \dfrac{\cos\theta}{\cos\theta}}$$
$$= \dfrac{1 + 2 \times \dfrac{1}{\tan\theta}}{\tan\theta - 2}.$$

Since $\tan\theta = 3$, it follows that
$$A = \dfrac{1 + 2 \times \dfrac{1}{3}}{3 - 2} = \dfrac{1 + \dfrac{2}{3}}{1} = \dfrac{5}{3}.$$

Therefore, $A = \dfrac{5}{3}$.

2. $B = \dfrac{\sin^2\theta + 4\cos^2\theta}{\sin^2\theta - 3\cos^2\theta}$;

We have $B = \dfrac{\dfrac{\sin^2\theta + 4\cos^2\theta}{\cos^2\theta}}{\dfrac{\sin^2\theta - 3\cos^2\theta}{\cos^2\theta}}$

$$= \frac{\dfrac{\sin^2\theta}{\cos^2\theta} + 4 \times \dfrac{\cos^2\theta}{\cos^2\theta}}{\dfrac{\sin^2\theta}{\cos^2\theta} - 3 \times \dfrac{\cos^2\theta}{\cos^2\theta}}$$

$$= \frac{\tan^2\theta + 4}{\tan^2\theta - 3}$$

$$= \frac{3^2 + 4}{3^2 - 3}$$

$$= \frac{9 + 4}{9 - 3}$$

$$= \frac{13}{6}.$$

Therefore, $B = \dfrac{13}{6}$.

3. $C = \dfrac{\sin^3\theta - \cos^3\theta + \cos\theta}{\sin^3\theta + 2\cos^3\theta - \sin\theta}$;

We have $C = \dfrac{\sin^3\theta - \cos^3\theta + \cos\theta}{\sin^3\theta + 2\cos^3\theta - \sin\theta}$

$$= \frac{\dfrac{\sin^3\theta - \cos^3\theta + \cos\theta}{\cos^3\theta}}{\dfrac{\sin^3\theta + 2\cos^3\theta - \sin\theta}{\cos^3\theta}}$$

$$= \frac{\dfrac{\sin^3\theta}{\cos^3\theta} - \dfrac{\cos^3\theta}{\cos^3\theta} + \dfrac{\cos\theta}{\cos^3\theta}}{\dfrac{\sin^3\theta}{\cos^3\theta} + 2 \times \dfrac{\cos^3\theta}{\cos^3\theta} - \dfrac{\sin\theta}{\cos^3\theta}}$$

$$= \frac{\tan^3\theta - 1 + \dfrac{1}{\cos^2\theta}}{\tan^3\theta + 2 - \dfrac{1}{\cos^2\theta}}.$$

Using the fact that $\dfrac{1}{\cos^2\theta} = 1 + \tan^2\theta$, we obtain

$$C = \frac{\tan^3\theta - 1 + 1 + \tan^2\theta}{\tan^3\theta + 2 - (1 + \tan^2\theta)}$$

$$= \frac{\tan^3\theta + \tan^2\theta}{\tan^3\theta - \tan^2\theta + 1}.$$

$$= \frac{3^3 + 3^2}{3^3 - 3^2 + 1}$$
$$= \frac{27 + 9}{27 - 9 + 1}$$
$$= \frac{36}{19}.$$

Therefore, $C = \dfrac{36}{19}$.

4. $D = \dfrac{\sin^4\theta - 2\sin^2\theta - 7}{\cos^4\theta + \cos^2\theta - 3}$.

We have $D = \dfrac{\sin^4\theta - 2\sin^2\theta - 7}{\cos^4\theta + \cos^2\theta - 3}$

$$= \frac{\dfrac{\sin^4\theta - 2\sin^2\theta - 7}{\cos^4\theta}}{\dfrac{\cos^4\theta + \cos^2\theta - 3}{\cos^4\theta}}$$

$$= \frac{\dfrac{\sin^4\theta}{\cos^4\theta} - 2 \times \dfrac{\sin^2\theta}{\cos^4\theta} - 7 \times \dfrac{1}{\cos^4\theta}}{\dfrac{\cos^4\theta}{\cos^4\theta} + \dfrac{\cos^2\theta}{\cos^4\theta} - 3 \times \dfrac{1}{\cos^4\theta}}$$

$$= \frac{\tan^4\theta - 2\tan^2\theta \left(\dfrac{1}{\cos^2\theta}\right) - 7\left(\dfrac{1}{\cos^2\theta}\right)^2}{1 + \dfrac{1}{\cos^2\theta} - 3\left(\dfrac{1}{\cos^2\theta}\right)^2}$$

$$= \frac{\tan^4\theta - 2\tan^2\theta\left(1 + \tan^2\theta\right) - 7\left(1 + \tan^2\theta\right)^2}{1 + 1 + \tan^2\theta - 3\left(1 + \tan^2\theta\right)^2}$$

$$= \frac{\tan^4\theta - 2\tan^2\theta\left(1 + \tan^2\theta\right) - 7\left(1 + \tan^2\theta\right)^2}{2 + \tan^2\theta - 3\left(1 + \tan^2\theta\right)^2}$$

$$= \frac{3^4 - 2 \times 3^2\left(1 + 3^2\right) - 7\left(1 + 3^2\right)^2}{2 + 3^2 - 3(1 + 3^2)^2}$$

$$= \frac{81 - 18 \times 10 - 7 \times 10^2}{2 + 9 - 3 \times 10^2}$$

$$= \frac{81 - 180 - 700}{11 - 300}$$

$$= \frac{-799}{-289}$$
$$= \frac{47}{17}.$$

Therefore, $D = \frac{47}{17}$.

Problem 8. Simplify the following expressions:

1. $\sqrt{\tan^2\theta + \cot^2\theta + 2}$;

2. $\dfrac{\sin^2\theta - \tan^2\theta}{\cos^2\theta - \cot^2\theta}$;

3. $\sqrt{\sin^2\theta\,(1+\cot\theta) + \cos^2\theta\,(1+\tan\theta)}$;

4. $\dfrac{\cos^2\theta + \cos^2\theta\cot^2\theta}{\sin^2\theta + \sin^2\theta\tan^2\theta}$;

5. $\dfrac{1+\cos 2\theta}{\sin 2\theta}$;

6. $\dfrac{2\sin\theta\sin 2\theta + \cos 3\theta + 3\cos\theta}{\sin\theta}$;

7. $\dfrac{2\cos 2\theta}{1+\cos 2\theta - \sin 2\theta}$.

Solution. Simplify the following expressions:

1. $\sqrt{\tan^2\theta + \cot^2\theta + 2}$

We have $\sqrt{\tan^2\theta + \cot^2\theta + 2}$
$$= \sqrt{\tan^2\theta + \frac{1}{\tan^2\theta} + 2}$$
$$= \sqrt{\tan^2\theta + 2\tan\theta\left(\frac{1}{\tan x}\right) + \frac{1}{\tan^2\theta}}$$
$$= \sqrt{\left(\tan\theta + \frac{1}{\tan\theta}\right)^2} = \left|\tan\theta + \frac{1}{\tan\theta}\right|$$
$$= \left|\frac{\sin\theta}{\cos\theta} + \frac{\cos\theta}{\sin\theta}\right| = \left|\frac{\sin^2\theta + \cos^2\theta}{\sin\theta\cos x}\right|$$

$$= \left|\frac{1}{\sin\theta\cos\theta}\right| = \frac{1}{|\sin\theta\cos\theta|}.$$

Thus, $\sqrt{\tan^2\theta + \cot^2\theta + 2} = \dfrac{1}{|\sin\theta\cos\theta|}$.

2. $\dfrac{\sin^2\theta - \tan^2\theta}{\cos^2\theta - \cot^2\theta}$

Observe that

$$\frac{\sin^2\theta - \tan^2\theta}{\cos^2\theta - \cot^2\theta} = \frac{\sin^2\theta - \dfrac{\sin^2\theta}{\cos^2\theta}}{\cos^2\theta - \dfrac{\cos^2\theta}{\sin^2\theta}}$$

$$= \frac{\dfrac{\sin^2\theta\cos^2\theta - \sin^2\theta}{\cos^2\theta}}{\dfrac{\cos^2\theta\sin^2\theta - \cos^2\theta}{\sin^2\theta}}$$

$$= \frac{\sin^2\theta\left(\cos^2\theta - 1\right)}{\cos^2\theta} \times \frac{\sin^2\theta}{\cos^2\theta\left(\sin^2\theta - 1\right)}$$

$$= \frac{\sin^4\theta\left(1 - \cos^2\theta\right)}{\cos^4\theta\left(1 - \sin^2\theta\right)}$$

$$= \frac{\sin^4\theta\left(\sin^2\theta\right)}{\cos^4\theta\left(\cos^2\theta\right)} = \frac{\sin^6\theta}{\cos^6\theta} = \tan^6\theta.$$

Therefore, $\dfrac{\sin^2\theta - \tan^2 x}{\cos^2\theta - \cot^2\theta} = \tan^6\theta.$

3. $\sqrt{\sin^2\theta\left(1 + \cot\theta\right) + \cos^2\theta\left(1 + \tan\theta\right)}$

We have $\sqrt{\sin^2\theta\left(1 + \cot\theta\right) + \cos^2\theta\left(1 + \tan\theta\right)}$

$$= \sqrt{\sin^2\theta\left(1 + \frac{\cos\theta}{\sin\theta}\right) + \cos^2\theta\left(1 + \frac{\sin\theta}{\cos\theta}\right)}$$

$$= \sqrt{\sin^2\theta + \sin\theta\cos x + \cos^2\theta + \sin\theta\cos\theta}$$

$$= \sqrt{\sin^2\theta + 2\sin\theta\cos\theta + \cos^2\theta}$$

$$= \sqrt{\left(\sin\theta + \cos\theta\right)^2} = |\sin\theta + \cos\theta|.$$

Therefore, $\sqrt{\sin^2\theta\left(1 + \cot\theta\right) + \cos^2\theta\left(1 + \tan\theta\right)} = |\sin\theta + \cos\theta|.$

4. $\dfrac{\cos^2\theta + \cos^2\theta\cot^2\theta}{\sin^2\theta + \sin^2\theta\tan^2\theta}$

We have $\dfrac{\cos^2\theta + \cos^2\theta\cot^2\theta}{\sin^2\theta + \sin^2\theta\tan^2\theta} = \dfrac{\cos^2\theta\left(1+\cot^2\theta\right)}{\sin^2\theta\left(1+\tan^2\theta\right)}$

$$= \dfrac{\cos^2\theta\left(\dfrac{1}{\sin^2\theta}\right)}{\sin^2\theta\left(\dfrac{1}{\cos^2\theta}\right)}$$

$$= \dfrac{\cot^2\theta}{\tan^2\theta} = \cot^4\theta.$$

Therefore, $\dfrac{\cos^2\theta + \cos^2\theta\cot^2\theta}{\sin^2\theta + \sin^2\theta\tan^2\theta} = \cot^4\theta.$

5. $\dfrac{1+\cos 2\theta}{\sin 2\theta}$

Using the fact that $\cos 2\theta = 2\cos^2\theta - 1$ and $\sin 2\theta = 2\sin\theta\cos\theta$, we obtain

$$\dfrac{1+\cos 2\theta}{\sin 2\theta} = \dfrac{1+2\cos^2\theta - 1}{2\sin\theta\cos\theta}$$

$$= \dfrac{2\cos^2\theta}{2\sin\theta\cos\theta}$$

$$= \dfrac{\cos\theta}{\sin\theta}$$

$$= \cot\theta.$$

Therefore, $\dfrac{1+\cos 2\theta}{\sin 2\theta} = \cot\theta.$

6. $\dfrac{2\sin\theta\sin 2\theta + \cos 3\theta + 3\cos\theta}{\sin\theta}$

By knowing that $\sin 2\theta = 2\sin\theta\cos\theta$ and $\cos 3\theta = 4\cos^3\theta - 3\cos\theta$, we obtain

$$\dfrac{2\sin\theta\sin 2\theta + \cos 3\theta + 3\cos\theta}{\sin\theta}$$

$$= \dfrac{2\sin\theta\left(2\sin\theta\cos\theta\right) + 4\cos^3\theta - 3\cos\theta + 3\cos\theta}{\sin\theta}$$

$$= \dfrac{4\sin^2\theta\cos\theta + 4\cos^3\theta}{\sin\theta}$$

$$= \frac{4\cos\theta\left(\sin^2\theta + \cos^2\theta\right)}{\sin\theta}$$
$$= \frac{4\cos\theta\,(1)}{\sin\theta}$$
$$= 4\cot\theta.$$

Therefore, $\dfrac{2\sin\theta\sin 2\theta + \cos 3\theta + 3\cos\theta}{\sin\theta} = 4\cot\theta.$

7. $\dfrac{2\cos 2\theta}{1 + \cos 2\theta - \sin 2\theta}$

We know that $\cos 2\theta = \cos^2\theta - \sin^2\theta$ and $\sin 2\theta = 2\sin\theta\cos\theta$. It follows that

$$\frac{2\cos 2\theta}{1 + \cos 2\theta - \sin 2\theta} = \frac{2\left(\cos^2\theta - \sin^2\theta\right)}{1 + 2\cos^2\theta - 1 - 2\sin\theta\cos\theta}$$
$$= \frac{2\left(\cos\theta - \sin\theta\right)\left(\cos\theta + \sin\theta\right)}{2\cos\theta\left(\cos\theta - \sin\theta\right)}$$
$$= \frac{\cos\theta + \sin\theta}{\cos\theta}$$
$$= \frac{\cos\theta}{\cos\theta} + \frac{\sin\theta}{\cos\theta}$$
$$= 1 + \tan\theta.$$

Therefore, $\dfrac{2\cos 2\theta}{1 + \cos 2\theta - \sin 2\theta} = 1 + \tan\theta.$

Problem 9. Prove the following equalities:

1. $\dfrac{\sin\theta + \cos\theta - 1}{\sin\theta - \cos\theta + 1} = \dfrac{\cos\theta}{1 + \sin\theta}$;

2. $\dfrac{\cos\theta\cot\theta - \sin\theta\tan\theta}{\dfrac{1}{\sin\theta} - \dfrac{1}{\cos\theta}} = 1 + \sin\theta\cos\theta$;

3. $(\tan\theta - \sin\theta)^2 + (1 - \cos\theta)^2 = \left(\dfrac{1}{\cos\theta} - 1\right)^2$;

4. $\dfrac{2\sin\theta\cos\theta - \cos\theta}{1 - \sin\theta + \sin^2\theta - \cos^2\theta} = \dfrac{1}{\tan\theta}$;

5. $\left(\dfrac{\sin^4\theta - \tan^4\theta}{\cos^4\theta - \cot^4\theta}\right)\left(\dfrac{\sin^2\theta + 1}{\cos^2\theta + 1}\right) = \tan^{10}\theta$;

6. $\dfrac{\cos 2\theta}{\sqrt{2}\cos\left(\theta + \dfrac{\pi}{4}\right)} = \cos\theta - \sin\theta.$

Solution. 1. $\dfrac{\sin\theta + \cos\theta - 1}{\sin\theta - \cos\theta + 1} = \dfrac{\cos\theta}{1 + \sin\theta}$
Observe that

$$(\sin\theta + \cos\theta - 1)(1 + \sin\theta)$$
$$= \sin\theta + \sin^2\theta + \cos\theta + \sin\theta\cos\theta - 1 - \sin\theta$$
$$= -1 + \sin^2\theta + \sin\theta\cos\theta + \cos\theta$$
$$= -\cos^2\theta + \sin\theta\cos\theta + \cos\theta = \cos\theta(\sin\theta - \cos\theta + 1).$$

It follows that $\dfrac{\sin\theta + \cos\theta - 1}{\sin\theta - \cos\theta + 1} = \dfrac{\cos\theta}{1 + \sin\theta}.$

2. $\dfrac{\cos\theta\cot\theta - \sin\theta\tan\theta}{\dfrac{1}{\sin\theta} - \dfrac{1}{\cos\theta}} = 1 + \sin\theta\cos\theta$

We have

$$\dfrac{\cos\theta\cot\theta - \sin\theta\tan\theta}{\dfrac{1}{\sin\theta} - \dfrac{1}{\cos\theta}}$$
$$= \dfrac{\cos\theta\left(\dfrac{\cos\theta}{\sin\theta}\right) - \sin\theta\left(\dfrac{\sin\theta}{\cos\theta}\right)}{\dfrac{\cos\theta - \sin\theta}{\sin\theta\cos\theta}}$$
$$= \left(\dfrac{\cos^2\theta}{\sin\theta} - \dfrac{\sin^2\theta}{\cos\theta}\right)\left(\dfrac{\sin\theta\cos\theta}{\cos\theta - \sin\theta}\right)$$
$$= \left(\dfrac{\cos^3\theta - \sin^3\theta}{\sin\theta\cos\theta}\right)\left(\dfrac{\sin\theta\cos\theta}{\cos\theta - \sin\theta}\right)$$
$$= \dfrac{(\cos\theta - \sin\theta)(\cos^2\theta + \sin\theta\cos\theta + \sin^2\theta)}{\cos\theta - \sin\theta}$$
$$= \sin^2\theta + \cos^2\theta + \sin\theta\cos\theta = 1 + \sin\theta\cos\theta.$$

Consequently, $\dfrac{\cos\theta\cot\theta - \sin\theta\tan\theta}{\dfrac{1}{\sin\theta} - \dfrac{1}{\cos\theta}} = 1 + \sin\theta\cos\theta.$

3. $(\tan\theta - \sin\theta)^2 + (1 - \cos\theta)^2 = \left(\dfrac{1}{\cos\theta} - 1\right)^2$

   We have $(\tan\theta - \sin\theta)^2 + (1 - \cos\theta)^2$
   $$= \left(\dfrac{\sin\theta}{\cos\theta} - \sin\theta\right)^2 + (1 - \cos\theta)^2$$
   $$= \sin^2\theta\left(\dfrac{1}{\cos\theta} - 1\right)^2 + \cos^2\theta\left(\dfrac{1}{\cos\theta} - 1\right)^2$$
   $$= \left(\dfrac{1}{\cos\theta} - 1\right)^2(\sin^2\theta + \cos^2\theta) = \left(\dfrac{1}{\cos\theta} - 1\right)^2.$$

   Consequently, $(\tan\theta - \sin\theta)^2 + (1 - \cos\theta)^2 = \left(\dfrac{1}{\cos\theta} - 1\right)^2.$

4. $\dfrac{2\sin\theta\cos\theta - \cos\theta}{1 - \sin\theta + \sin^2\theta - \cos^2\theta} = \dfrac{1}{\tan\theta}$

   We have
   $$\dfrac{2\sin\theta\cos\theta - \cos\theta}{1 - \sin\theta + \sin^2\theta - \cos^2\theta} = \dfrac{\cos\theta\,(2\sin\theta - 1)}{1 - \cos^2\theta - \sin\theta + \sin^2\theta}$$
   $$= \dfrac{\cos\theta\,(2\sin\theta - 1)}{\sin^2\theta - \sin\theta + \sin^2\theta}$$
   $$= \dfrac{\cos\theta\,(2\sin\theta - 1)}{2\sin^2\theta - \sin\theta}$$
   $$= \dfrac{\cos\theta\,(2\sin\theta - 1)}{\sin\theta\,(2\sin\theta - 1)}$$
   $$= \dfrac{\cos\theta}{\sin\theta} = \cot\theta = \dfrac{1}{\tan\theta}.$$

   Consequently, $\dfrac{2\sin\theta\cos\theta - \cos\theta}{1 - \sin\theta + \sin^2\theta - \cos^2\theta} = \dfrac{1}{\tan\theta}.$

5. $\left(\dfrac{\sin^4\theta - \tan^4\theta}{\cos^4\theta - \cot^4\theta}\right)\left(\dfrac{\sin^2\theta + 1}{\cos^2\theta + 1}\right) = \tan^{10}\theta$

   We know that $\tan\theta = \dfrac{\sin\theta}{\cos\theta}$ and $\cot\theta = \dfrac{\cos\theta}{\sin\theta}$.

   Hence, $\dfrac{\sin^4\theta - \tan^4\theta}{\cos^4\theta - \cot^4\theta} = \dfrac{\sin^4\theta - \left(\dfrac{\sin\theta}{\cos\theta}\right)^4}{\cos^4\theta - \left(\dfrac{\cos\theta}{\sin\theta}\right)^4}$

$$= \frac{\sin^4\theta - \dfrac{\sin^4\theta}{\cos^4\theta}}{\cos^4\theta - \dfrac{\cos^4\theta}{\sin^4\theta}}$$

$$= \frac{\sin^4\theta\left(1 - \dfrac{1}{\cos^4\theta}\right)}{\cos^4\theta\left(1 - \dfrac{1}{\sin^4\theta}\right)}$$

$$= \tan^4\theta \times \frac{\dfrac{\cos^4\theta - 1}{\cos^4\theta}}{\dfrac{\sin^4\theta - 1}{\sin^4\theta}}$$

$$= \tan^4\theta \times \frac{\sin^4\theta}{\cos^4\theta} \times \frac{\cos^4\theta - 1}{\sin^4\theta - 1}$$

$$= \tan^4\theta \times \tan^4\theta \times \frac{(\cos^2\theta - 1)(\cos^2\theta + 1)}{(\sin^2\theta - 1)(\sin^2\theta + 1)}$$

$$= \tan^8\theta \times \frac{(-\sin^2\theta)(\cos^2\theta + 1)}{(-\cos^2\theta)(\sin^2\theta + 1)}$$

$$= \tan^8\theta \times \tan^2\theta \times \left(\frac{\cos^2\theta + 1}{\sin^2\theta + 1}\right)$$

$$= \tan^{10}\theta \times \left(\frac{\cos^2\theta + 1}{\sin^2\theta + 1}\right).$$

Multiply both sides of the equality by $\dfrac{\sin^2\theta + 1}{\cos^2\theta + 1}$, we obtain

$$\left(\frac{\sin^4\theta - \tan^4\theta}{\cos^4\theta - \cot^4\theta}\right)\left(\frac{\sin^2\theta + 1}{\cos^2\theta + 1}\right) = \tan^{10}\theta.$$

6. $\dfrac{\cos 2\theta}{\sqrt{2}\cos\left(\theta + \dfrac{\pi}{4}\right)} = \cos\theta - \sin\theta$

Using the fact that $\cos 2\theta = \cos^2\theta - \sin^2\theta$ and $\cos(a + b) = \cos a \cos b - \sin a \sin b$, we obtain

$$\frac{\cos 2\theta}{\sqrt{2}\cos\left(\theta + \dfrac{\pi}{4}\right)} = \frac{\cos^2\theta - \sin^2\theta}{\sqrt{2}\left(\cos\theta\cos\dfrac{\pi}{4} + \sin\theta\sin\dfrac{\pi}{4}\right)}$$

$$= \frac{(\cos\theta - \sin\theta)(\cos\theta + \sin\theta)}{\sqrt{2}\left(\dfrac{\sqrt{2}}{2}\cos\theta + \dfrac{\sqrt{2}}{2}\sin\theta\right)}$$

$$= \frac{(\cos\theta - \sin\theta)(\cos\theta + \sin\theta)}{\dfrac{\sqrt{2^2}}{2}\cos\theta + \dfrac{\sqrt{2^2}}{2}\sin\theta}$$

$$= \frac{(\cos\theta - \sin\theta)(\cos\theta + \sin\theta)}{\cos\theta + \sin\theta}$$

$$= \cos\theta - \sin\theta.$$

Therefore, $\dfrac{\cos 2\theta}{\sqrt{2}\cos\left(\theta + \dfrac{\pi}{4}\right)} = \cos\theta - \sin\theta.$

Problem 10. Prove the following equalities:

1. $\dfrac{\sin^4\theta + 2\sin\theta\cos\theta - \cos^4\theta}{\tan 2\theta - 1} = \cos 2\theta;$

2. $\dfrac{\cos^3\theta - \cos 3\theta}{\cos\theta} + \dfrac{\sin^3\theta + \sin 3\theta}{\sin\theta} = 3;$

3. $\dfrac{\sin 3\theta}{\sin^3\theta} + \dfrac{\cos 3\theta}{\cos^3\theta} = \dfrac{12\cot 2\theta}{\sin 2\theta}.$

Solution. Prove the following equalities:

1. $\dfrac{\sin^4\theta + 2\sin\theta\cos\theta - \cos^4\theta}{\tan 2\theta - 1} = \cos 2\theta$
We have

$$\frac{\sin^4\theta + 2\sin\theta\cos\theta - \cos^4\theta}{\tan 2\theta - 1}$$

$$= \frac{\sin^4\theta - \cos^4\theta + 2\sin\theta\cos\theta}{\dfrac{\sin 2\theta}{\cos 2\theta} - 1}$$

$$= \frac{(\sin^2\theta - \cos^2\theta)(\sin^2\theta + \cos^2\theta) + \sin 2\theta}{\dfrac{\sin 2\theta - \cos 2\theta}{\cos 2\theta}}$$

$$= \left(\frac{-\cos 2\theta + \sin 2\theta}{\sin 2\theta - \cos 2\theta}\right)\cos 2\theta$$

$$= \cos 2\theta.$$

Therefore, $\dfrac{\sin^4\theta + 2\sin\theta\cos\theta - \cos^4\theta}{\tan 2\theta - 1} = \cos 2\theta$.

2. $\dfrac{\cos^3\theta - \cos 3\theta}{\cos\theta} + \dfrac{\sin^3\theta + \sin 3\theta}{\sin\theta} = 3$

Since $\sin 3\theta = 3\sin\alpha - 4\sin^3\theta$ and $\cos 3\theta = 4\cos^3\theta - 3\cos\theta$, we obtain

$\dfrac{\cos^3\theta - \cos 3\theta}{\cos\theta} + \dfrac{\sin^3\theta + \sin 3\theta}{\sin\theta}$

$= \dfrac{\cos^3\theta - (4\cos^3\theta - 3\cos\theta)}{\cos\theta} + \dfrac{\sin^3\theta + (3\sin\theta - 4\sin^3\theta)}{\sin\theta}$

$= \dfrac{-3\cos^3\theta + 3\cos\theta}{\cos\theta} + \dfrac{-3\sin^3\theta + 3\sin\theta}{\cos\theta}$

$= \dfrac{(-3\cos^2\theta + 3)\cos\theta}{\cos\theta} + \dfrac{(-3\sin^2\theta + 3)\sin\theta}{\sin\theta}$

$= -3\cos^2\theta + 3 - 3\sin^2\theta + 3$

$= -3(\cos^2\theta + \sin^2\theta) + 6$

$= -3 + 6 = 3$.

Hence, $\dfrac{\cos^3\theta - \cos 3\theta}{\cos\theta} + \dfrac{\sin^3\theta + \sin 3\theta}{\sin\theta} = 3$.

3. $\dfrac{\sin 3\theta}{\sin^3\theta} + \dfrac{\cos 3\theta}{\cos^3\theta} = \dfrac{12\cot 2\theta}{\sin 2\theta}$

We know that $\sin 3\theta = 3\sin\theta - 4\sin^3\theta$ and $\cos 3\theta = 3\cos\theta - 4\cos^3\theta$. Hence,

$\dfrac{\sin 3\theta}{\sin^3\theta} + \dfrac{\cos 3\theta}{\cos^3\theta} = \dfrac{3\sin\theta - 4\sin^3\theta}{\sin^3\theta} + \dfrac{4\cos^3\theta - 3\cos\theta}{\cos^3\theta}$

$= \dfrac{3\sin\theta}{\sin^3\theta} - \dfrac{4\sin^3\theta}{\sin^3\theta} + \dfrac{4\cos^3\theta}{\cos^3\theta} - \dfrac{3\cos\theta}{\cos^3\theta}$

$= 3\left(\dfrac{1}{\sin^2\theta}\right) - 4 + 4 - 3\left(\dfrac{1}{\cos^2\theta}\right)$

$= 3\left(\dfrac{1}{\sin^2\theta} - \dfrac{1}{\cos^2\theta}\right)$

$= 3\left(\dfrac{\cos^2\theta - \sin^2\theta}{\sin^2\theta\cos^2\theta}\right)$.

Chapter 8. Solutions

Since $\cos 2\theta = \cos^2\theta - \sin^2\theta$ and $\sin 2\theta = 2\sin\theta\cos\theta$, we obtain

$$\frac{\sin 3\theta}{\sin^3\theta} + \frac{\cos 3\theta}{\cos^3\theta} = 12\left(\frac{\cos 2\theta}{\sin^2 2\theta}\right)$$

$$= 12 \times \frac{\cos 2\theta}{\sin 2\theta} \times \frac{1}{\sin 2\theta}$$

$$= 12 \times \cot 2\theta \times \frac{1}{\sin 2\theta}$$

$$= \frac{12\cot 2\theta}{\sin 2\theta}.$$

Problem 11. Prove the following equalities:

1. $\sin 3\theta = 4\sin\theta \sin\left(\dfrac{\pi}{3} + \theta\right)\sin\left(\dfrac{\pi}{3} - \theta\right);$

2. $\cos 3\theta = 4\cos\theta \cos\left(\dfrac{\pi}{3} + \theta\right)\cos\left(\dfrac{\pi}{3} - \theta\right);$

3. $4\sin\left(\theta + \dfrac{\pi}{3}\right)\sin\left(\theta - \dfrac{\pi}{3}\right) = 4\sin^2\theta - 3;$

4. $\sin(\theta - \gamma)\sin(\theta + \gamma)\cos(2\theta - 2\gamma) + \cos(4\theta - 2\gamma) + \cos 2\gamma$
$\qquad = \cos(2\theta - 4\gamma) + \cos 2\theta.$

Solution.   1. $\sin 3\theta = 4\sin\theta\sin\left(\dfrac{\pi}{3} + \theta\right)\sin\left(\dfrac{\pi}{3} - \theta\right)$
Observe that

$4\sin\theta\sin\left(\dfrac{\pi}{3} + \theta\right)\sin\left(\dfrac{\pi}{3} - \theta\right)$

$= 2\left[2\sin\left(\dfrac{\pi}{3} + \theta\right)\sin\left(\dfrac{\pi}{3} - \theta\right)\right]\sin\theta$

$= 2\left[\cos\left(\dfrac{\pi}{3} + \theta - \dfrac{\pi}{3} + \theta\right) - \cos\left(\dfrac{\pi}{3} + \theta + \dfrac{\pi}{3} - \theta\right)\right]\sin\theta$

$= 2\left(\cos 2\theta - \cos\dfrac{2\pi}{3}\right)\sin\theta$

$= 2\left(1 - 2\sin^2\theta + \cos\dfrac{\pi}{3}\right) = 2\left(1 - 2\sin^2\theta + \dfrac{1}{2}\right)\sin\theta$

$= 2\left(\dfrac{3}{2} - 2\sin^2\theta\right)\sin\theta = 3\sin\theta - 4\sin^3\theta = \sin 3\theta.$

Consequently, $\sin 3\theta = 4\sin\theta\sin\left(\dfrac{\pi}{3} + \theta\right)\sin\left(\dfrac{\pi}{3} - \theta\right).$

2. $\cos 3\theta = 4\cos\theta \cos\left(\dfrac{\pi}{3}+\theta\right)\cos\left(\dfrac{\pi}{3}-\theta\right)$
We have

$$4\cos\theta \cos\left(\dfrac{\pi}{3}+\theta\right)\cos\left(\dfrac{\pi}{3}-\theta\right)$$
$$= 2\left[2\cos\left(\dfrac{\pi}{3}+\theta\right)\cos\left(\dfrac{\pi}{3}-\theta\right)\right]\cos\theta$$
$$= 2\left[\cos\left(\dfrac{\pi}{3}+\theta-\dfrac{\pi}{3}+\theta\right)+\cos\left(\dfrac{\pi}{3}+\theta+\dfrac{\pi}{3}-\theta\right)\right]\cos\theta$$
$$= 2\left(\cos 2\theta + \cos\dfrac{2\pi}{3}\right)\cos\theta = 2\left(2\cos^2\theta - 1 - \dfrac{1}{2}\right)\cos\theta$$
$$= 2\left(2\cos^2\theta - \dfrac{3}{2}\right)\cos\theta$$
$$= 4\cos^3\theta - 3\cos\theta = \cos 3\theta.$$

Thus, $\cos 3\theta = 4\cos\theta\cos\left(\dfrac{\pi}{3}+\theta\right)\cos\left(\dfrac{\pi}{3}-\theta\right)$.

3. $4\sin\left(\theta+\dfrac{\pi}{3}\right)\sin\left(\theta-\dfrac{\pi}{3}\right) = 4\sin^2\theta - 3$
We have

$$4\sin\left(\theta+\dfrac{\pi}{3}\right)\sin\left(\theta-\dfrac{\pi}{3}\right)$$
$$= 2\left[\cos\left(\theta+\dfrac{\pi}{3}-\theta+\dfrac{\pi}{3}\right) - \cos\left(\theta+\dfrac{\pi}{3}+\theta-\dfrac{\pi}{3}\right)\right]$$
$$= 2\left(\cos\dfrac{2\pi}{3} - \cos 2\theta\right)$$
$$= 2\left[-\dfrac{1}{2} - (1 - 2\sin^2\theta)\right]$$
$$= 2\left(-\dfrac{3}{2} + 2\sin^2\theta\right)$$
$$= -3 + 4\sin^2\theta = 4\sin^2\theta - 3.$$

Therefore, $4\sin\left(\theta+\dfrac{\pi}{3}\right)\sin\left(\theta-\dfrac{\pi}{3}\right) = 4\sin^2\theta - 3$.

4. $4\sin(\theta-\gamma)\sin(\theta+\gamma)\cos(2\theta-2\gamma) + \cos(4\theta-2\gamma) + \cos 2\gamma$
$$= \cos(2\theta-4\gamma) + \cos 2\theta.$$
Using product to sum formula, we obtain

$4\sin(\theta-\gamma)\sin(\theta+\gamma)\cos(2\theta-2\gamma)$

$$= 2\left[2\sin\left(\theta-\gamma\right)\sin\left(\theta+\gamma\right)\right]\cos\left(2\theta-2\gamma\right)$$
$$= 2\left[\cos\left(\theta-\gamma-\theta-\gamma\right)-\cos\left(\theta-\gamma+\theta+\gamma\right)\right]\cos\left(2\theta-2\gamma\right)$$
$$= 2\left(\cos 2\gamma - \cos 2\theta\right)\cos\left(2\theta-2\gamma\right)$$
$$= 2\cos 2\gamma \cos\left(2\theta-2\gamma\right) - 2\cos 2\theta \cos\left(2\theta-2\gamma\right)$$
$$= \cos\left(2\gamma - 2\theta + 2\gamma\right) + \cos\left(2\gamma + 2\theta - 2\gamma\right) - \cos\left(2\theta - 2\theta + 2\gamma\right)$$
$$\quad - \cos\left(2\theta + 2\theta - 2\gamma\right)$$
$$= \cos\left(4\gamma - 2\theta\right) + \cos 2\theta - \cos 2\gamma - \cos\left(4\theta - 2\gamma\right)$$
$$= \cos\left(2\theta - 4\gamma\right) + \cos 2\theta - \cos 2\gamma - \cos\left(4\theta - 2\gamma\right).$$

Therefore, $4\sin\left(\theta-\gamma\right)\sin\left(\theta+\gamma\right)\cos\left(2\theta-2\gamma\right) + \cos\left(4\theta-2\gamma\right)$
$+ \cos 2\gamma = \cos\left(2\theta - 4\gamma\right) + \cos 2\theta.$

Problem 12. For all integers k, prove that

1. $\sin\left(k\pi - \alpha\right) = (-1)^{k+1}\sin\alpha$;

2. $\cos\left(k\pi - \alpha\right) = (-1)^{k}\cos\alpha$;

3. $\tan\left(k\pi - \alpha\right) = -\tan\alpha$;

4. $\cot\left(k\pi - \alpha\right) = -\cot\alpha$.

Solution. Prove that

1. $\sin\left(k\pi - \alpha\right) = (-1)^{k+1}\sin\alpha$
   Observe that $\sin\left(\pi - \alpha\right) = \sin\alpha = (-1)^{1+1}\sin\alpha$.
   Suppose that $\sin\left(k\pi - \alpha\right) = (-1)^{k+1}\sin\alpha$.
   We shall show that $\sin\left[(k+1)\pi - \alpha\right] = (-1)^{k+2}\sin\alpha$.
   We have

$$\sin\left[(k+1)\pi - \alpha\right] = \sin\left(k\pi + \pi - \alpha\right)$$
$$= -\sin\left(k\pi - \alpha\right)$$
$$= (-1)(-1)^{k+1}\sin\alpha$$
$$= (-1)^{k+2}\sin\alpha.$$

Consequently, $\sin\left(k\pi - \alpha\right) = (-1)^{k+1}\sin\alpha$ for all $k \in \mathbb{N}$.
Let $k = -t$, where $t \in \mathbb{N}$.
We obtain

$$\sin\left(k\pi - \alpha\right) = \sin\left(-t\pi - \alpha\right)$$

$$= -\sin\left[t\pi - (-\alpha)\right]$$
$$= -(-1)^{t+1}\sin(-\alpha)$$
$$= (-1)^{-k+1}\sin\alpha$$
$$= (-1)^{-k-1}\sin\alpha = (-1)^{k+1}\sin\alpha.$$

Then $\sin(k\pi - \alpha) = (-1)^{k+1}\sin\alpha$ for all negative integers k. Therefore, $\sin(k\pi - \alpha) = (-1)^{k+1}\sin\alpha$ for all $k \in \mathbb{Z}$.

2. $\cos(k\pi - \alpha) = (-1)^{k+1}\cos\alpha$
   Observe that $\cos(\pi - \alpha) = -\cos\alpha = (-1)^1\cos\alpha$.
   Suppose that $\cos(k\pi - \alpha) = (-1)^k\cos\alpha$.
   We shall show that $\cos[(k+1)\pi - \alpha] = (-1)^{k+1}\cos\alpha$.
   We have
   $$\cos[(k+1)\pi - \alpha] = \cos(\pi + k\pi - \alpha)$$
   $$= -\cos(k\pi - \alpha)$$
   $$= -(-1)^k\cos\alpha$$
   $$= (-1)^{k+1}\cos\alpha.$$

Then $\cos(k\pi - \alpha) = (-1)^k\cos\alpha$ for all $k \in \mathbb{N}$.
Let $k = -t$, where $t \in \mathbb{N}$.
It follows that
$$\cos(k\pi - \alpha) = \cos(-t\pi - \alpha)$$
$$= \cos(t\pi + \alpha)$$
$$= \cos[t\pi - (-\alpha)]$$
$$= (-1)^t\cos(-\alpha)$$
$$= (-1)^{-k}\cos\alpha = (-1)^k\cos\alpha.$$

We obtain $\cos(k\pi - \alpha) = (-1)^k\cos\alpha$ for all negative integers k. Therefore, the claim is proved.

3. $\tan(k\pi - \alpha) = -\tan\alpha$
   Since $\tan x = \dfrac{\sin x}{\cos x}$, it follows that
   $$\tan(k\pi - \alpha) = \frac{\sin(k\pi - \alpha)}{\cos(k\pi - \alpha)} = \frac{(-1)^{k+1}\sin\alpha}{(-1)^k\cos\alpha} = -\tan\alpha.$$

   Consequently, $\tan(k\pi - \alpha) = -\tan\alpha$.

4. $\cot(k\pi - \alpha) = -\cot\alpha$

Using $\cot x = \dfrac{1}{\tan x}$, we obtain

$$\cot(k\pi - \alpha) = \frac{1}{\tan(k\pi - \alpha)}$$
$$= \frac{1}{-\tan\alpha}$$
$$= -\frac{1}{\tan\alpha}$$
$$= -\cot\alpha.$$

Therefore, $\cot(k\pi - \alpha) = -\cot\alpha$.

Problem 13. Let A, B and C be the three angles of a triangle. For all integers k, prove the following identities:

1. $\sin kA = (-1)^{k+1} \sin k(B + C)$;
2. $\cos kA = (-1)^{k} \cos k(B + C)$;
3. $\tan kA = -\tan k(B + C)$;
4. $\cot kA = -\cot k(B + C)$.

Solution. Since A, B and C are the three angles of a triangle, then

$$A + B + C = \pi.$$

Prove that:

1. $\sin kA = (-1)^{k+1} \sin k(B + C)$

   We have

   $$\sin kA = \sin k[\pi - (B + C)]$$
   $$= \sin[k\pi - k(B + C)]$$
   $$= (-1)^{k+1} \sin k(B + C)$$

   because $\sin(k\pi - \alpha) = (-1)^{k+1} \sin\alpha$.
   Consequently, $\sin kA = (-1)^{k+1} \sin k(B + C)$.

2. $\cos kA = (-1)^{k+1} \cos k(B + C)$

   We have

   $$\cos kA = \cos k[\pi - (B + C)]$$

$$= \cos\left[k\pi - k\left(B+C\right)\right]$$
$$= (-1)^k \cos k\left(B+C\right)$$

because $\cos(k\pi - \alpha) = (-1)^k \cos\alpha$.
Hence, $\cos kA = (-1)^k \cos k\left(B+C\right)$.

3. $\tan kA = -\tan k\left(B+C\right)$
Observe that

$$\tan kA = \tan k\left[\pi - (B+C)\right]$$
$$= \tan\left[k\pi - k\left(B+C\right)\right]$$
$$= -\tan k\left(B+C\right)$$

because $\tan(k\pi - \alpha) = -\tan\alpha$.
Therefore, $\tan kA = -\tan k\left(B+C\right)$.

4. $\cot kA = -\cot k\left(B+C\right)$
We have

$$\cot kA = \cot k\left[\pi - (B+C)\right]$$
$$= \cot\left[k\pi - k\left(B+C\right)\right]$$
$$= -\cot k\left(B+C\right)$$

because $\cot(k\pi - \alpha) = -\cot\alpha$.
Thus, $\cot kA = -\cot k\left(B+C\right)$.

Problem 14. Let ABC be a triangle. Prove the following identities:

1. $\sin(2k+1)\dfrac{A}{2} = (-1)^k \cos(2k+1)\left(\dfrac{B+C}{2}\right)$;

2. $\cos(2k+1)\dfrac{A}{2} = (-1)^k \sin(2k+1)\left(\dfrac{B+C}{2}\right)$;

3. $\tan(2k+1)\dfrac{A}{2} = \cot(2k+1)\dfrac{B}{2}$;

4. $\cot(2k+1)\dfrac{A}{2} = \tan(2k+1)\left(\dfrac{B+C}{2}\right)$.

Solution. Since ABC is a triangle, it follows that $A+B+C = \pi$. Prove that:

101

Chapter 8. Solutions

1. $\sin(2k+1)\dfrac{A}{2} = (-1)^k \cos(2k+1)\left(\dfrac{B+C}{2}\right)$

Using $\sin\left(\dfrac{\pi}{2}+\alpha\right) = \cos\alpha$ and $\cos(k\pi - \alpha) = (-1)^k \cos\alpha$, we obtain

$$\sin(2k+1)\dfrac{A}{2} = \sin(2k+1)\left[\dfrac{\pi - (B+C)}{2}\right]$$

$$= \sin\left[(2k+1)\dfrac{\pi}{2} - (2k+1)\left(\dfrac{B+C}{2}\right)\right]$$

$$= \sin\left[\dfrac{\pi}{2} + k\pi - (2k+1)\left(\dfrac{B+C}{2}\right)\right]$$

$$= \cos\left[k\pi - (2k+1)\left(\dfrac{B+C}{2}\right)\right]$$

$$= (-1)^k \cos(2k+1)\left(\dfrac{B+C}{2}\right).$$

Therefore, $\sin(2k+1)\dfrac{A}{2} = (-1)^k \cos(2k+1)\left(\dfrac{B+C}{2}\right)$.

2. $\cos(2k+1)\dfrac{A}{2} = (-1)^k \sin(2k+1)\left(\dfrac{B+C}{2}\right)$

Using $\cos\left(\dfrac{\pi}{2}+\alpha\right) = -\sin\alpha$ and $\sin(k\pi - \alpha) = (-1)^{k+1}\sin\alpha$, it follows that

$$\cos(2k+1)\dfrac{A}{2} = \cos(2k+1)\left[\dfrac{\pi - (B+C)}{2}\right]$$

$$= \cos\left[(2k+1)\dfrac{\pi}{2} - (2k+1)\left(\dfrac{B+C}{2}\right)\right]$$

$$= \cos\left[\dfrac{\pi}{2} + k\pi - (2k+1)\left(\dfrac{B+C}{2}\right)\right]$$

$$= -\sin\left[k\pi - (2k+1)\left(\dfrac{B+C}{2}\right)\right]$$

$$= -(-1)^{k+1}\sin(2k+1)\left(\dfrac{B+C}{2}\right)$$

$$= (-1)^k \sin(2k+1)\left(\dfrac{B+C}{2}\right).$$

Therefore, $\cos(2k+1)\dfrac{A}{2} = (-1)^k \sin(2k+1)\left(\dfrac{B+C}{2}\right) =$

$$(-1)^k \sin(2k+1)\left(\frac{B+C}{2}\right).$$

3. $\tan(2k+1)\frac{A}{2} = \cot(2k+1)\frac{B}{2}$

We have

$$\tan(2k+1)\frac{A}{2} = \frac{\sin(2k+1)\frac{A}{2}}{\cos(2k+1)\frac{A}{2}}$$

$$= \frac{(-1)^k \cos(2k+1)\left(\frac{B+C}{2}\right)}{(-1)^k \sin(2k+1)\left(\frac{B+C}{2}\right)}$$

$$= \cot(2k+1)\left(\frac{B+C}{2}\right).$$

Therefore, $\tan(2k+1)\frac{A}{2} = \cot(2k+1)\frac{B}{2}$.

4. $\cot(2k+1)\frac{A}{2} = \tan(2k+1)\left(\frac{B+C}{2}\right)$.

Using $\cot x = \frac{\cos x}{\sin x}$, it implies that

$$\cot(2k+1)\frac{A}{2} = \frac{\cos(2k+1)\frac{A}{2}}{\sin(2k+1)\frac{A}{2}}$$

$$= \frac{(-1)^k \sin(2k+1)\left(\frac{B+C}{2}\right)}{(-1)^k \cos(2k+1)\left(\frac{B+C}{2}\right)}$$

$$= \tan(2k+1)\left(\frac{B+C}{2}\right).$$

Therefore, $\cot(2k+1)\frac{A}{2} = \tan(2k+1)\left(\frac{B+C}{2}\right)$.

Problem 15. Let ABC be a triangle. Prove the following equalities:

1. $\sin A + \sin B + \sin C = 4\cos\frac{A}{2}\cos\frac{B}{2}\cos\frac{C}{2}$;
2. $\sin 2A + \sin 2B + \sin 2C = 4\sin A \sin B \sin C$;
3. $\sin 3A + \sin 3B + \sin 3C = -4\cos\frac{3A}{2}\cos\frac{3B}{2}\cos\frac{3C}{2}$;

4. $\sin 4A + \sin 4B + \sin 4C = -4\sin 2A \sin 2B \sin 2C$;

5. $\sin(2k+1)A + \sin(2k+1)B + \sin(2k+1)C$;
$$= (-1)^k \cos(2k+1)\frac{A}{2} \cos(2k+1)\frac{B}{2} \cos(2k+1)\frac{C}{2}$$

6. $\sin 2kA + \sin 2kB + \sin 2kC = 4(-1)^{k+1} \sin kA \sin kB \sin kC$.

Solution.   1. $\sin A + \sin B + \sin C = 4\cos\frac{A}{2}\cos\frac{B}{2}\cos\frac{C}{2}$
We have

$\sin A + \sin B + \sin C$
$$= 2\sin\left(\frac{A+B}{2}\right)\cos\left(\frac{A-B}{2}\right) + 2\sin\frac{C}{2}\cos\frac{C}{2}$$
$$= 2\sin\left(\frac{\pi-C}{2}\right)\cos\left(\frac{A-B}{2}\right) + 2\sin\left[\frac{\pi-(A+B)}{2}\right]\cos\frac{C}{2}$$
$$= 2\cos\frac{C}{2}\cos\left(\frac{A-B}{2}\right) + 2\cos\left(\frac{A+B}{2}\right)\cos\frac{C}{2}$$
$$= 2\left[\cos\left(\frac{A-B}{2}\right) + \cos\left(\frac{A+B}{2}\right)\right]\cos\frac{C}{2}$$
$$= 4\cos\frac{A}{2}\cos\frac{B}{2}\cos\frac{C}{2}.$$

Consequently, $\sin A + \sin B + \sin C = 4\cos\frac{A}{2}\cos\frac{B}{2}\cos\frac{C}{2}$.

2. $\sin 2A + \sin 2B + \sin 2C = 4\sin A \sin B \sin C$
Consider

$\sin 2A + \sin 2B + \sin 2C$
$$= 2\sin(A+B)\cos(A-B) + 2\sin C \cos C$$
$$= 2\sin(\pi-C)\cos(A-B) + 2\sin C \cos[\pi-(A+B)]$$
$$= 2\sin C \cos(A-B) - 2\sin C \cos(A+B)$$
$$= 2\sin C[\cos(A-B) - \cos(A+B)]$$
$$= -4\sin C \sin A \sin(-B).$$

Hence, $\sin 2A + \sin 2B + \sin 2C = 4\sin A \sin B \sin C$.

3. $\sin 3A + \sin 3B + \sin 3C = -4\cos\dfrac{3A}{2}\cos\dfrac{3B}{2}\cos\dfrac{3C}{2}$
We have

$\sin 3A + \sin 3B + \sin 3C$
$= 2\sin\left(\dfrac{3A+3B}{2}\right)\cos\left(\dfrac{3A-3B}{2}\right) + 2\sin\dfrac{3C}{2}\cos\dfrac{3C}{2}$
$= 2\sin\left(\dfrac{3\pi-3C}{2}\right)\cos\left(\dfrac{3A-3B}{2}\right) + 2\sin\dfrac{3C}{2}\cos\dfrac{3C}{2}$
$= -2\cos\dfrac{3C}{2}\cos\left(\dfrac{3A-3B}{2}\right) + 2\sin\dfrac{3C}{2}\cos\dfrac{3C}{2}$
$= -2\left[\cos\left(\dfrac{3A-3B}{2}\right) - \sin\dfrac{3C}{2}\right]\cos\dfrac{3C}{2}$
$= -2\left[\cos\left(\dfrac{3A-3B}{2}\right) + \cos\left(\dfrac{3A+3B}{2}\right)\right]\cos\dfrac{3C}{2}$
$= -4\cos\dfrac{3A}{2}\cos\dfrac{3B}{2}\cos\dfrac{3C}{2}.$

Therefore, $\sin 3A + \sin 3B + \sin 3C = -4\cos\dfrac{3A}{2}\cos\dfrac{3B}{2}\cos\dfrac{3C}{2}$.

4. $\sin 4A + \sin 4B + \sin 4C = -4\sin 2A \sin 2B \sin 2C$
We have

$\sin 4A + \sin 4B + \sin 4C$
$= 2\sin(2A+2B)\cos(2A-2B) + 2\sin 2C \cos 2C$
$= 2\sin(2\pi - 2C)\cos(2A-2B) + 2\sin 2C \cos 2C$
$= -2\sin 2C \cos(2A-2B) + 2\sin 2C \cos 2C$
$= -2\sin 2C[\cos(2A-2B) - \cos 2C]$
$= -2\sin 2C[\cos(2A-2B) - \cos(2A+2B)]$
$= 4\sin 2C \sin 2A \sin(-2B).$

Therefore, $\sin 4A + \sin 4B + \sin 4C = -4\sin 2A \sin 2B \sin 2C$.

5. $\sin(2k+1)A + \sin(2k+1)B + \sin(2k+1)C$
$= (-1)^k \cos(2k+1)\dfrac{A}{2}\cos(2k+1)\dfrac{B}{2}\cos(2k+1)\dfrac{C}{2}$

Using sum to product formula, we obtain

$\sin(2k+1)A + \sin(2k+1)B + \sin(2k+1)C$
$= 2\sin\left[\dfrac{(2k+1)A+(2k+1)B}{2}\right]\cos\left[\dfrac{(2k+1)A-(2k+1)B}{2}\right]$
$\quad + 2\sin(2k+1)\dfrac{C}{2}\cos(2k+1)\dfrac{C}{2}$
$= 2\sin(2k+1)\left(\dfrac{A+B}{2}\right)\cos(2k+1)\left(\dfrac{A-B}{2}\right)$
$\quad + 2\sin(2k+1)\dfrac{C}{2}\cos(2k+1)\dfrac{C}{2}$
$= 2(-1)^k\cos(2k+1)\dfrac{C}{2}\cos(2k+1)\left(\dfrac{A-B}{2}\right)$
$\quad + 2(-1)^k\cos(2k+1)\left(\dfrac{A+B}{2}\right)\cos(2k+1)\dfrac{C}{2}$
$= 2(-1)^k\cos(2k+1)\dfrac{C}{2}$
$\quad \times \left[\cos(2k+1)\left(\dfrac{A-B}{2}\right) + \cos(2k+1)\left(\dfrac{A+B}{2}\right)\right]$
$= 4(-1)^k\cos(2k+1)\dfrac{C}{2}\cos\left[\dfrac{(2k+1)\left(\frac{A-B}{2}\right)-(2k+1)\left(\frac{A+B}{2}\right)}{2}\right]$
$\quad \times \cos\left[\dfrac{(2k+1)\left(\frac{A-B}{2}\right)+(2k+1)\left(\frac{A+B}{2}\right)}{2}\right]$
$= 4(-1)^k\cos(2k+1)\dfrac{C}{2}\cos(2k+1)\left(\dfrac{A-B-A-B}{4}\right)$
$\quad \times \cos(2k+1)\left(\dfrac{A-B+A+B}{4}\right)$
$= 4(-1)^k\cos(2k+1)\dfrac{C}{2}\cos(2k+1)\dfrac{A}{2}\cos(2k+1)\dfrac{B}{2}.$

Consequently, $\sin(2k+1)A + \sin(2k+1)B + \sin(2k+1)C$
$= (-1)^k\cos(2k+1)\dfrac{A}{2}\cos(2k+1)\dfrac{B}{2}\cos(2k+1)\dfrac{C}{2}.$

6. $\sin 2kA + \sin 2kB + \sin 2kC = 4(-1)^{k+1}\sin kA\sin kB\sin kC.$

We have

$$\sin 2kA + \sin 2kB + \sin 2kC$$
$$= 2\sin\left(\frac{2kA+2kB}{2}\right)\cos\left(\frac{2kA-2kB}{2}\right) + 2\sin kC \cos kC$$
$$= 2\sin(A+B)k \cos(A-B)k + 2\sin kC \cos kC$$
$$= 2\sin(\pi - C)k \cos(A-B)k + 2\sin kC \cos kC$$
$$= 2\sin(k\pi - kC) \cos(A-B)k + 2\sin kC \cos kC$$
$$= 2(-1)^{k+1}\sin kC \cos(A-B)k + 2\sin kC \cos kC$$
$$= 2(-1)^{k+1}\sin kC \cos(A-B)k + 2(-1)^k \sin kC \cos(A+B)k$$
$$= 2(-1)^{k+1}\sin kC [\cos(A-B)k - \cos(A+B)k]$$
$$= -4(-1)^{k+1}\sin kC \sin\left[\frac{(A-B)k - (A+B)k}{2}\right]$$
$$\times \sin\left[\frac{(A-B)k + (A+B)k}{2}\right]$$
$$= -4(-1)^{k+1}\sin kC \sin(-Bk)\sin Ak$$
$$= 4(-1)^{k+1}\sin Ak \sin Bk \sin Ck.$$

Therefore, $\sin 2kA + \sin 2kB + \sin 2kC$
$$= 4(-1)^{k+1}\sin kA \sin kB \sin kC.$$

Problem 16. Let ABC be a triangle. Prove the following equalities:

1. $\cos A + \cos B + \cos C = 1 + 4\sin\frac{A}{2}\sin\frac{B}{2}\sin\frac{C}{2}$;
2. $\cos 2A + \cos 2B + \cos 2C = -1 - 4\cos A \cos B \cos C$;
3. $\cos 3A + \cos 3B + \cos 3C = -1 - 4\sin\frac{3A}{2}\sin\frac{3B}{2}\sin\frac{3C}{2}$;
4. $\cos 4A + \cos 4B + \cos 4C = -1 + 4\cos 2A \cos 2B \cos 2C$;
5. $\cos(2k+1)A + \cos(2k+1)B + \cos(2k+1)C$
$$= 1 + 4(-1)^k \sin(2k+1)\frac{A}{2}\sin(2k+1)\frac{B}{2}\sin(2k+1)\frac{C}{2};$$
6. $\cos 2kA + \cos 2kB + \cos 2kC = -1 + 4(-1)^k \cos kA \cos kB \cos kC$.

Chapter 8. Solutions

Solution. Prove the following equalities:

1. $\cos A + \cos B + \cos C = 1 + 4\sin\dfrac{A}{2}\sin\dfrac{B}{2}\sin\dfrac{C}{2}$

Consider

$\cos A + \cos B + \cos C$

$= 2\cos\left(\dfrac{A+B}{2}\right)\cos\left(\dfrac{A-B}{2}\right) + 1 - 2\sin^2\dfrac{C}{2}$

$= 1 + 2\sin\dfrac{C}{2}\cos\left(\dfrac{A-B}{2}\right) - 2\sin^2\dfrac{C}{2}$

$= 1 + 2\sin\dfrac{C}{2}\left[\cos\left(\dfrac{A-B}{2}\right) - \sin\dfrac{C}{2}\right]$

$= 1 + 2\sin\dfrac{C}{2}\left[\cos\left(\dfrac{A-B}{2}\right) - \cos\left(\dfrac{A+B}{2}\right)\right]$

$= 1 - 4\sin\dfrac{C}{2}\sin\dfrac{A}{2}\sin\left(\dfrac{-B}{2}\right).$

Thus, $\cos A + \cos B + \cos C = 1 + 4\sin\dfrac{A}{2}\sin\dfrac{B}{2}\sin\dfrac{C}{2}$.

2. $\cos 2A + \cos 2B + \cos 2C = -1 - 4\cos A\cos B\cos C$

We have

$\cos 2A + \cos 2B + \cos 2C$
$= 2\cos(A+B)\cos(A-B) + 2\cos^2 C - 1$
$= -1 - 2\cos C\cos(A-B) + 2\cos^2 C$
$= -1 - 2\cos C[\cos(A-B) - \cos C]$
$= -1 - 2\cos C[\cos(A-B) + \cos(A+B)]$
$= -1 - 4\cos A\cos B\cos C.$

Consequently, $\cos 2A + \cos 2B + \cos 2C = -1 - 4\cos A\cos B\cos C$.

3. $\cos 3A + \cos 3B + \cos 3C = 1 - 4\sin\dfrac{3A}{2}\sin\dfrac{3B}{2}\sin\dfrac{3C}{2}$

We have

$\cos 3A + \cos 3B + \cos 3C$

$= 2\cos\left(\dfrac{3A+3B}{2}\right)\cos\left(\dfrac{3A-3B}{2}\right) + 1 - 2\sin^2\dfrac{3C}{2}$

108

$$= 2\cos\left(\frac{3\pi - 3C}{2}\right)\cos\left(\frac{3A - 3B}{2}\right) + 1 - 2\sin^2\frac{3C}{2}$$

$$= 1 - 2\sin\frac{3C}{2}\cos\left(\frac{3A - 3B}{2}\right) - 2\sin^2\frac{3C}{2}$$

$$= 1 - 2\sin\frac{3C}{2}\left[\cos\left(\frac{3A - 3B}{2}\right) - \cos\left(\frac{3A + 3B}{2}\right)\right]$$

$$= 1 - 4\sin\frac{3A}{2}\sin\frac{3B}{2}\sin\frac{3C}{2}.$$

Consequently, $\cos 3A + \cos 3B + \cos 3C$

$$= 1 - 4\sin\frac{3A}{2}\sin\frac{3B}{2}\sin\frac{3C}{2}$$

4. $\cos 4A + \cos 4B + \cos 4C = -1 + 4\cos 2A \cos 2B \cos 2C$

We have

$\cos 4A + \cos 4B + \cos 4C$

$$= 2\cos\left(\frac{4A + 4B}{2}\right)\cos\left(\frac{4A - 4B}{2}\right) + 2\cos^2 2C - 1$$

$$= 2\cos 2(A + B)\cos 2(A - B) + 2\cos^2 2C - 1$$

$$= 2\cos 2(\pi - C)\cos 2(A - B) + 2\cos^2 2C - 1$$

$$= 2\cos(2\pi - 2C)\cos 2(A - B) + 2\cos^2 2C - 1$$

$$= 2\cos 2C \cos 2(A - B) + 2\cos^2 2C - 1$$

$$= 2\cos 2C \left[\cos 2(A - B) + \cos 2C\right] - 1$$

$$= 4\cos 2C \cos\left[\frac{2(A - B) + 2C}{2}\right]\cos\left[\frac{2(A - B) - 2C}{2}\right] - 1$$

$$= 4\cos 2C \cos(A - B + C)\cos(A - B - C) - 1$$

$$= 4\cos 2C \cos(\pi - 2B)\cos(2A - \pi) - 1$$

$$= -1 + 4\cos 2A \cos 2B \cos 2C.$$

Therefore, $\cos 4A + \cos 4B + \cos 4C = -1 + 4\cos 2A \cos 2B \cos 2C.$

5. $\cos(2k+1)A + \cos(2k+1)B + \cos(2k+1)C$

$$= 1 + 4(-1)^k \sin(2k+1)\frac{A}{2}\sin(2k+1)\frac{B}{2}\sin(2k+1)\frac{C}{2}$$

We have

$\cos(2k+1)A + \cos(2k+1)B + \cos(2k+1)C$

Chapter 8. Solutions

$$= 2\cos\left[\frac{(2k+1)A + (2k+1)B}{2}\right]\cos\left[\frac{(2k+1)A - (2k+1)B}{2}\right]$$
$$+ 1 - 2\sin^2(2k+1)\frac{C}{2}$$
$$= 1 + 2\cos(2k+1)\left(\frac{A+B}{2}\right)\cos(2k+1)\left(\frac{A-B}{2}\right)$$
$$- 2\sin^2(2k+1)\frac{C}{2}$$
$$= 1 + 2(-1)^k \sin(2k+1)\frac{C}{2}\cos(2k+1)\left(\frac{A-B}{2}\right)$$
$$- 2\sin^2(2k+1)\frac{C}{2}$$
$$= 1 + 2(-1)^k \sin(2k+1)\frac{C}{2}$$
$$\times \left[\cos(2k+1)\left(\frac{A-B}{2}\right) - (-1)^k \sin(2k+1)\frac{C}{2}\right]$$
$$= 1 + 2(-1)^k \sin(2k+1)\frac{C}{2}$$
$$\times \left[\cos(2k+1)\left(\frac{A-B}{2}\right) - \cos(2k+1)\left(\frac{A+B}{2}\right)\right]$$
$$= 1 -$$
$$4(-1)^k \sin(2k+1)\frac{C}{2}\sin\left[\frac{(2k+1)\left(\frac{A-B}{2}\right) - (2k+1)\left(\frac{A+B}{2}\right)}{2}\right]$$
$$\times \sin\left[\frac{(2k+1)\left(\frac{A-B}{2}\right) + (2k+1)\left(\frac{A+B}{2}\right)}{2}\right]$$
$$= 1 - 4(-1)^k \sin(2k+1)\frac{C}{2}\sin(2k+1)\left(\frac{A-B-A-B}{4}\right)$$
$$\times \sin(2k+1)\left(\frac{A-B+A+B}{4}\right)$$
$$= 1 + 4(-1)^k \sin(2k+1)\frac{C}{2}\sin(2k+1)\frac{B}{2}\sin(2k+1)\frac{A}{2}.$$

Thus, the claim is proved.

6. $\cos 2kA + \cos 2kB + \cos 2kC = -1 + 4(-1)^k \cos kA \cos kB \cos kC$

We have
$$\cos 2kA + \cos 2kB + \cos 2kC$$

$$= 2\cos\left(\frac{2kA+2kB}{2}\right)\cos\left(\frac{2kA-2kB}{2}\right) + 2\cos^2 kC - 1$$
$$= 2\cos(A+B)k\cos(A-B)k + 2\cos^2 kC - 1$$
$$= 2(-1)^k \cos kC \cos(A-B)k + 2\cos^2 kC - 1$$
$$= 2(-1)^k \cos kC \left[\cos(A-B)k + (-1)^k \cos kC\right] - 1$$
$$= 2(-1)^k \cos kC \left[\cos(A-B)k + \cos(A+B)k\right] - 1$$
$$= -1 + 4(-1)^k \cos kC \cos\left[\frac{(A-B)k + (A+B)k}{2}\right]$$
$$\times \cos\left[\frac{(A-B)k - (A+B)k}{2}\right]$$
$$= -1 + 4(-1)^k \cos kC \cos kA \cos kB.$$

Consequently,
$$\cos 2kA + \cos 2kB + \cos 2kC = -1 + 4(-1)^k \cos kA \cos kB \cos kC.$$

Problem 17. Let ABC be a triangle. Prove the following equalities:

1. $\tan A + \tan B + \tan C = \tan A \tan B \tan C$;
2. $\tan 2A + \tan 2B + \tan 2C = \tan 2A \tan 2B \tan 2C$;
3. $\tan kA + \tan kB + \tan kC = \tan kA \tan kB \tan kC$;
4. $\tan \dfrac{A}{2} \tan \dfrac{B}{2} + \tan \dfrac{B}{2} \tan \dfrac{C}{2} + \tan \dfrac{C}{2} \tan \dfrac{A}{2} = 1$;
5. $\tan(2k+1)\dfrac{A}{2}\tan(2k+1)\dfrac{B}{2} + \tan(2k+1)\dfrac{B}{2}\tan(2k+1)\dfrac{C}{2}$
$$+ \tan(2k+1)\dfrac{C}{2}\tan(2k+1)\dfrac{A}{2} = 1.$$

Solution. Since ABC is a triangle, $A + B + C = \pi$.
Prove the following equalities:

1. $\tan A + \tan B + \tan C = \tan A \tan B \tan C$
   We have
   $$\tan(A+B+C) = \frac{\tan(A+B) + \tan C}{1 - \tan(A+B)\tan C}$$

Chapter 8. Solutions

$$= \frac{\dfrac{\tan A + \tan B}{1 - \tan A \tan B} + \tan C}{1 - \left(\dfrac{\tan A + \tan B}{1 - \tan A \tan B}\right) \tan C}$$

$$= \frac{\tan A + \tan B + \tan C - \tan A \tan B \tan C}{1 - \tan A \tan B - \tan B \tan C - \tan C \tan A}.$$

By knowing that $\tan(A + B + C) = \tan \pi = 0$, we obtain

$$\frac{\tan A + \tan B + \tan C - \tan A \tan B \tan C}{1 - \tan A \tan B - \tan B \tan C - \tan C \tan A} = 0.$$

Consequently, $\tan A + \tan B + \tan C = \tan A \tan B \tan C$.

2. $\tan 2A + \tan 2B + \tan 2C = \tan 2A \tan 2B \tan 2C$
Using the same proof in (1), we obtain

$$\tan(2A + 2B + 2C)$$
$$= \frac{\tan 2A + \tan 2B + \tan 2C - \tan 2A \tan 2B \tan 2C}{1 - \tan 2A \tan 2B - \tan 2B \tan 2C - \tan 2C \tan 2A}.$$

Since $\tan(2A + 2B + 2C) = \tan 2\pi = 0$, it implies that

$$\frac{\tan 2A + \tan 2B + \tan 2C - \tan 2A \tan 2B \tan 2C}{1 - \tan 2A \tan 2B - \tan 2B \tan 2C - \tan 2C \tan 2A} = 0.$$

Hence, $\tan 2A + \tan 2B + \tan 2C = \tan 2A \tan 2B \tan 2C$.

3. $\tan kA + \tan kB + \tan kC = \tan kA \tan kB \tan kC$
We have

$$\tan(kA + kB + kC)$$
$$= \frac{\tan kA + \tan kB + \tan kC - \tan kA \tan kB \tan kC}{1 - \tan kA \tan kB - \tan kB \tan kC - \tan kC \tan kA}.$$

By $\tan(kA + kB + kC) = \tan k(A + B + C) = \tan k\pi = 0$, it follows that

$$\frac{\tan kA + \tan kB + \tan kC - \tan kA \tan kB \tan kC}{1 - \tan kA \tan kB - \tan kB \tan kC - \tan kC \tan kA} = 0.$$

Hence, $\tan kA + \tan kB + \tan kC = \tan kA \tan kB \tan kC$.

4. $\tan\dfrac{A}{2}\tan\dfrac{B}{2} + \tan\dfrac{B}{2}\tan\dfrac{C}{2} + \tan\dfrac{C}{2}\tan\dfrac{A}{2} = 1$

We have $\dfrac{A}{2} + \dfrac{B}{2} = \dfrac{\pi}{2} - \dfrac{C}{2}$, it follows that

$$\tan\left(\dfrac{A}{2} + \dfrac{B}{2}\right) = \tan\left(\dfrac{\pi}{2} - \dfrac{C}{2}\right).$$

Then

$$\dfrac{\tan\dfrac{A}{2} + \tan\dfrac{B}{2}}{1 - \tan\dfrac{A}{2}\tan\dfrac{B}{2}} = \dfrac{1}{\tan\dfrac{C}{2}}$$

or

$$\tan\dfrac{A}{2}\tan\dfrac{C}{2} + \tan\dfrac{B}{2}\tan\dfrac{C}{2} = 1 - \tan\dfrac{A}{2}\tan\dfrac{B}{2}.$$

Consequently, $\tan\dfrac{A}{2}\tan\dfrac{B}{2} + \tan\dfrac{B}{2}\tan\dfrac{C}{2} + \tan\dfrac{C}{2}\tan\dfrac{A}{2} = 1$.

5. $\tan(2k+1)\dfrac{A}{2}\tan(2k+1)\dfrac{B}{2} + \tan(2k+1)\dfrac{B}{2}\tan(2k+1)\dfrac{C}{2}$

$\qquad + \tan(2k+1)\dfrac{C}{2}\tan(2k+1)\dfrac{A}{2} = 1$

We have $(2k+1)\dfrac{A}{2} + (2k+1)\dfrac{B}{2} + (2k+1)\dfrac{C}{2}$

$= (2k+1)\left(\dfrac{A+B+C}{2}\right)$

$= (2k+1)\dfrac{\pi}{2}.$

Then $(2k+1)\dfrac{A}{2} + (2k+1)\dfrac{B}{2} = (2k+1)\dfrac{\pi}{2} - (2k+1)\dfrac{C}{2}.$

We obtain

$$\tan\left[(2k+1)\dfrac{A}{2} + (2k+1)\dfrac{B}{2}\right] = \dfrac{1}{\tan(2k+1)\dfrac{C}{2}}$$

or

$$\dfrac{\tan(2k+1)\dfrac{A}{2} + \tan(2k+1)\dfrac{B}{2}}{1 - \tan(2k+1)\dfrac{A}{2}\tan(2k+1)\dfrac{B}{2}} = \dfrac{1}{\tan(2k+1)\dfrac{C}{2}}.$$

Chapter 8. Solutions

It follows that

$$\tan(2k+1)\frac{A}{2}\tan(2k+1)\frac{C}{2} + \tan(2k+1)\frac{B}{2}\tan(2k+1)\frac{C}{2}$$
$$= 1 - \tan(2k+1)\frac{A}{2}\tan(2k+1)\frac{B}{2}.$$

Therefore,

$$\tan(2k+1)\frac{A}{2}\tan(2k+1)\frac{B}{2} + \tan(2k+1)\frac{B}{2}\tan(2k+1)\frac{C}{2}$$
$$+ \tan(2k+1)\frac{C}{2}\tan(2k+1)\frac{A}{2} = 1.$$

Problem 18. Let ABC be a triangle. Prove the following equalities:

1. $\cot A \cot B + \cot B \cot C + \cot C \cot A = 1$;
2. $\cot kA \cot kB + \cot kB \cot kC + \cot kC \cot kA = 1$;
3. $\cot \frac{A}{2} + \cot \frac{B}{2} + \cot \frac{C}{2} = \cot \frac{A}{2} \cot \frac{B}{2} \cot \frac{C}{2}$;
4. $\cot(2k+1)\frac{A}{2} + \cot(2k+1)\frac{B}{2} + \cot(2k+1)\frac{C}{2}$
$$= \cot(2k+1)\frac{A}{2} \cot(2k+1)\frac{B}{2} \cot(2k+1)\frac{C}{2}.$$

Solution. Since ABC is a triangle, then $A + B + C = \pi$.
Prove the following equalities:

1. $\cot A \cot B + \cot B \cot C + \cot C \cot A = 1$
    Using addition formula, $\tan(p+q) = \dfrac{\tan p + \tan q}{1 - \tan p \tan q}$.
    Then
    $$\frac{1}{\tan(p+q)} = \frac{1 - \tan p \tan q}{\tan p + \tan q}$$
    or
    $$\cot(p+q) = \frac{1 - \dfrac{1}{\cot p \cot q}}{\dfrac{1}{\cot p} + \dfrac{1}{\cot q}} = \frac{\cot p \cot q - 1}{\cot p + \cot q}.$$

114

It follows that $\cot(A+B) = \dfrac{\cot A \cot B - 1}{\cot A + \cot B}$.
From $A + B = \pi - C$, we obtain
$$\cot(A+B) = \cot(\pi - C) = -\cot C.$$
Hence,
$$\dfrac{\cot A \cot B - 1}{\cot A + \cot B} = -\cot C$$
or
$$\cot A \cot B - 1 = -\cot A \cot C - \cot B \cot C.$$

Hence, $\cot A \cot B + \cot B \cot C + \cot C \cot A = 1$.

2. $\cot kA \cot kB + \cot kB \cot kC + \cot kC \cot kA = 1$
Observe that $kA + kB = k\pi - kC$.
It implies that
$$\cot(kA + kB) = \cot(k\pi - kC)$$
or
$$\dfrac{\cot kA \cot kB - 1}{\cot kA + \cot kB} = -\cot kC.$$
We obtain $\cot kA \cot kB - 1 = -\cot kA \cot kC - \cot kB \cot kC$.
Consequently,
$$\cot kA \cot kB + \cot kB \cot kC + \cot kC \cot kA = 1.$$

3. $\cot \dfrac{A}{2} + \cot \dfrac{B}{2} + \cot \dfrac{C}{2} = \cot \dfrac{A}{2} \cot \dfrac{B}{2} \cot \dfrac{C}{2}$
We have $\dfrac{A}{2} + \dfrac{B}{2} = \dfrac{\pi}{2} - \dfrac{C}{2}$.
It follows that
$$\cot\left(\dfrac{A}{2} + \dfrac{B}{2}\right) = \cot\left(\dfrac{\pi}{2} - \dfrac{C}{2}\right)$$
or
$$\dfrac{\cot \dfrac{A}{2} \cot \dfrac{B}{2} - 1}{\cot \dfrac{A}{2} + \cot \dfrac{B}{2}} = \dfrac{1}{\cot \dfrac{C}{2}}.$$
Then $\cot \dfrac{A}{2} \cot \dfrac{B}{2} \cot \dfrac{C}{2} - \cot \dfrac{C}{2} = \cot \dfrac{A}{2} + \cot \dfrac{B}{2}$.
Therefore, $\cot \dfrac{A}{2} + \cot \dfrac{B}{2} + \cot \dfrac{C}{2} = \cot \dfrac{A}{2} \cot \dfrac{B}{2} \cot \dfrac{C}{2}$.

4. $\cot(2k+1)\dfrac{A}{2} + \cot(2k+1)\dfrac{B}{2} + \cot(2k+1)\dfrac{C}{2}$

$= \cot(2k+1)\dfrac{A}{2} \cot(2k+1)\dfrac{B}{2} \cot(2k+1)\dfrac{C}{2}$

Observe that $(2k+1)\dfrac{A}{2} + (2k+1)\dfrac{B}{2} = (2k+1)\dfrac{\pi}{2} - (2k+1)\dfrac{C}{2}$.
Then

$\cot\left[(2k+1)\dfrac{A}{2} + (2k+1)\dfrac{B}{2}\right] = \cot\left[(2k+1)\dfrac{\pi}{2} - (2k+1)\dfrac{C}{2}\right]$

or

$\dfrac{\cot(2k+1)\dfrac{A}{2} \cot(2k+1)\dfrac{B}{2} - 1}{\cot(2k+1)\dfrac{A}{2} + \cot(2k+1)\dfrac{B}{2}} = \dfrac{1}{\cot(2k+1)\dfrac{C}{2}}.$

We obtain

$\cot(2k+1)\dfrac{A}{2} \cot(2k+1)\dfrac{B}{2} \cot(2k+1)\dfrac{C}{2} - \cot(2k+1)\dfrac{C}{2}$

$= \cot(2k+1)\dfrac{A}{2} + \cot(2k+1)\dfrac{B}{2}.$

Consequently,

$\cot(2k+1)\dfrac{A}{2} + \cot(2k+1)\dfrac{B}{2} + \cot(2k+1)\dfrac{C}{2}$
$= \cot(2k+1)\dfrac{A}{2} \cot(2k+1)\dfrac{B}{2} \cot(2k+1)\dfrac{C}{2}.$

Problem 19. Let ABC be a triangle. Show that

1. $\sin^3 A \cos(B-C) + \sin^3 B \cos(C-A) + \sin^3 C \cos(A-B) = 3 \sin A \sin B \sin C$;

2. $\sin^3 A \sin(B-C) + \sin^3 B \sin(C-A) + \sin^3 C \sin(A-B) = 0$;

3. $\sin 3A \sin^3(B-C) + \sin 3B \sin^3(C-A) + \sin 3C \sin^3(A-B) = 0$;

4. $\sin 3A\cos^3(B-C) + \sin 3B\cos^3(C-A) + \sin 3C\cos^3(A-B) = \sin 3A \sin 3B \sin 3C$.

Solution. Since ABC is a triangle, then $A+B+C = \pi$.
Show that

1. $\sin^3 A \cos(B-C) + \sin^3 B \cos(C-A) + \sin^3 C \cos(A-B) = 3\sin A \sin B \sin C$

Observe that

$$\sin^3 A \cos(B-C) = \sin^2 A [\sin A \cos(B-C)]$$
$$= \sin^2 A [\sin(B+C) \cos(B-C)].$$

Using product to sum formula, we obtain

$$\sin^3 A \cos(B-C)$$
$$= \frac{1}{2}\sin^2 A [\sin(B+C+B-C) + \sin(B+C-B+C)]$$
$$= \frac{1}{2}\sin^2 A (\sin 2B + \sin 2C)$$
$$= \frac{1}{2}\sin^2 A (2\sin B \cos B + 2\sin C \cos C)$$
$$= \sin^2 A (\sin B \cos B + \sin C \cos C).$$

Similarly,

$$\sin^3 B \cos(C-A) = \sin^2 B (\sin C \cos C + \sin A \cos A)$$

and

$$\sin^3 C \cos(A-B) = \sin^2 C (\sin A \cos A + \sin B \cos B).$$

It follows that

$$\sin^3 A \cos(B-C) + \sin^3 B \cos(C-A) + \sin^3 C \cos(A-B)$$
$$= \sin^2 A (\sin B \cos B + \sin C \cos C)$$
$$+ \sin^2 B (\sin C \cos C + \sin A \cos A)$$
$$+ \sin^2 C (\sin A \cos A + \sin B \cos B)$$
$$= \sin^2 A \sin B \cos B + \sin^2 A \sin C \cos C + \sin^2 B \sin C \cos C$$
$$+ \sin A \sin^2 B \cos A + \sin B \sin^2 C \cos B + \sin A \sin^2 C \cos A$$

Chapter 8. Solutions

$$\begin{aligned}
&= \left(\sin^2 A \sin B \cos B + \sin A \sin^2 B \cos A\right) \\
&+ \left(\sin^2 B \sin C \cos C + \sin B \sin^2 C \cos B\right) \\
&+ \left(\sin^2 A \sin C \cos C + \sin A \sin^2 C \cos A\right) \\
&= \sin A \sin B \left(\sin A \cos B + \sin B \cos A\right) \\
&+ \sin B \sin C \left(\sin B \cos C + \sin C \cos B\right) \\
&+ \sin C \sin A \left(\sin A \cos C + \sin C \cos A\right) \\
&= \sin A \sin B \sin (A+B) + \sin B \sin C \sin (B+C) \\
&+ \sin C \sin A \sin (C+A) \\
&= \sin A \sin B \sin C + \sin A \sin B \sin C + \sin A \sin B \sin C \\
&= 3 \sin A \sin B \sin C.
\end{aligned}$$

2. $\sin^3 A \sin(B-C) + \sin^3 B \sin(C-A) + \sin^3 C \sin(A-B) = 0$
   We have

$$\begin{aligned}
&\sin^3 A \sin (B-C) \\
&= \sin^2 A \left[\sin A \sin (B-C)\right] \\
&= \frac{1}{2}\sin^2 A \left[\sin(B+C)\sin(B-C)\right] \\
&= \frac{1}{2}\sin^2 A \left[\cos(B+C-B+C) - \cos(B+C+B-C)\right] \\
&= \frac{1}{2}\sin^2 A \left(\cos 2C - \cos 2B\right) \\
&= \frac{1}{2}\sin^2 A \left(1 - 2\sin^2 C - 1 + 2\sin^2 B\right) \\
&= \sin^2 A \left(\sin^2 B - \sin^2 C\right).
\end{aligned}$$

Likewise,

$$\sin^3 B \sin(C-A) = \sin^2 B \left(\sin^2 C - \sin^2 A\right)$$

and

$$\sin^3 C \sin(A-B) = \sin^2 C \left(\sin^2 A - \sin^2 B\right).$$

Hence,

$$\begin{aligned}
&\sin^3 A \sin(B-C) + \sin^3 B \sin(C-A) + \sin^3 C \sin(A-B) \\
&= \sin^2 A \left(\sin^2 B - \sin^2 C\right) + \sin^2 B \left(\sin^2 C - \sin^2 A\right)
\end{aligned}$$

$+\sin^2 C \left(\sin^2 A - \sin^2 B\right)$
$= \sin^2 A\sin^2 B - \sin^2 A\sin^2 C + \sin^2 B\sin^2 C - \sin^2 A\sin^2 B$
$+ \sin^2 A\sin^2 C - \sin^2 B\sin^2 C$
$= 0.$

3. $\sin 3A\sin^3(B-C) + \sin 3B\sin^3(C-A) + \sin 3C\sin^3(A-B) = 0.$
Using triple formula, $\sin 3x = 3\sin x - 4\sin^3 x$.
Then $\sin^3 x = \dfrac{3}{4}\sin x - \dfrac{1}{4}\sin 3x.$
It follows that

$\sin 3A\sin^3(B-C)$
$= \sin 3A \left[\dfrac{3}{4}\sin(B-C) - \dfrac{1}{4}\sin 3(B-C)\right]$
$= \dfrac{3}{4}\sin 3A \sin(B-C) - \dfrac{1}{4}\sin 3A\sin(3B-3C)$
$= \dfrac{3}{8}[\cos(3A-B+C) - \cos(3A+B-C)]$
$\quad - \dfrac{1}{8}[\cos(3A-3B+3C) - \cos(3A+3B-3C)]$
$= \dfrac{3}{8}[\cos 2(A-C) - \cos 2(A-B)] - \dfrac{1}{8}(\cos 6C - \cos 6B).$

Similarly, we obtain

$\sin 3B\sin^3(C-A)$
$= \dfrac{3}{8}[\cos 2(A-B) - \cos 2(B-C)] - \dfrac{1}{8}(\cos 6A - \cos 6C)$

and

$\sin 3C\sin^3(A-B)$
$= \dfrac{3}{8}[\cos 2(B-C) - \cos 2(A-C)] - \dfrac{1}{8}(\cos 6B - \cos 6A).$

Adding the last three equalities, it implies that

$\sin 3A\sin^3(B-C) + \sin 3B\sin^3(C-A) + \sin 3C\sin^3(A-B) = 0.$

4. $\sin 3A\cos^3(B-C) + \sin 3B\cos^3(C-A) + \sin 3C\cos^3(A-B) =$
$\sin 3A \sin 3B \sin 3C.$

Chapter 8. Solutions

We have $\cos 3x = 4\cos^3 x - 3\cos x$.
Then $\cos^3 x = \dfrac{3}{4}\cos x + \dfrac{1}{4}\cos 3x$.
Hence,

$\sin 3A \cos^3(B-C)$
$= \sin 3A \left[\dfrac{3}{4}\cos(B-C) + \dfrac{1}{4}\cos 3(B-C)\right]$
$= \dfrac{3}{4}\sin 3A \cos(B-C) + \dfrac{1}{4}\sin 3A \cos(3B-3C)$
$= \dfrac{3}{8}[\sin(3A+B-C) + \sin(3A-B+C)]$
$\quad + \dfrac{1}{8}[\sin(3A+3B-3C) + \sin(3A-3B+3C)]$
$= \dfrac{3}{8}[\sin 2(C-A) + \sin 2(B-A)] + \dfrac{1}{8}(\sin 6C + \sin 6B)$.

Similarly,

$\sin 3B \cos^3(C-A)$
$= \dfrac{3}{8}[\sin 2(A-B) + \sin 2(C-B)] + \dfrac{1}{8}(\sin 6A + 6C)$

and

$\sin 3C \cos^3(A-B)$
$= \dfrac{3}{8}[\sin 2(B-C) + \sin 2(A-C)] + \dfrac{1}{8}(\sin 6A + 6B)$.

Adding the last three equalities, we obtain
$\sin 3A \cos^3(B-C) + \sin 3B \cos^3(C-A) + \sin 3C \cos^3(A-B)$
$= \dfrac{1}{8}(2\sin 6A + 2\sin 6B + 2\sin 6C)$
$= \dfrac{1}{4}(\sin 6A + \sin 6B + \sin 6C)$
$= \dfrac{1}{4}(4\sin 3A \sin 3B \sin 3C)$
$= \sin 3A \sin 3B \sin 3C$.

Problem 20. Let x, y and z be three real numbers such that
$$x + y + z = 0.$$

Prove the following equalities:
1. $\sin kx + \sin ky + \sin kz = -4\sin\dfrac{kx}{2}\sin\dfrac{ky}{2}\sin\dfrac{kz}{2}$;
2. $\cos kx + \cos ky + \cos kz = 4\cos\dfrac{kx}{2}\cos\dfrac{ky}{2}\cos\dfrac{kz}{2} - 1$;
3. $\tan kx + \tan ky + \tan kz = \tan kx \tan ky \tan kz$;
4. $\cot kx \cot ky + \cot ky \cot kz + \cot kz \cot kx = 1$.

Solution. Prove the following equalities:
1. $\sin kx + \sin ky + \sin kz = -4\sin\dfrac{kx}{2}\sin\dfrac{ky}{2}\sin\dfrac{kz}{2}$;
Since $x + y + z = 0$, it follows that $kx + ky + kz = 0$.
Using sum to product and double angle formulas, we obtain

$\sin kx + \sin ky + \sin kz$

$= 2\sin\left(\dfrac{kx+ky}{2}\right)\cos\left(\dfrac{kx-ky}{2}\right) + 2\sin\dfrac{kz}{2}\cos\dfrac{kz}{2}$

$= 2\sin\left(\dfrac{-kz}{2}\right)\cos\left(\dfrac{kx-ky}{2}\right) + 2\sin\dfrac{kz}{2}\cos\dfrac{kz}{2}$

$= -2\sin\dfrac{kz}{2}\cos\left(\dfrac{kx-ky}{2}\right) + 2\sin\dfrac{kz}{2}\cos\dfrac{kz}{2}$

$= -2\sin\dfrac{kz}{2}\left[\cos\left(\dfrac{kx-ky}{2}\right) - \cos\dfrac{kz}{2}\right]$

$= -2\sin\dfrac{kz}{2}(-2)\sin\left(\dfrac{\frac{kx-ky}{2}+\frac{kz}{2}}{2}\right)\sin\left(\dfrac{\frac{kx-ky}{2}-\frac{kz}{2}}{2}\right)$

$= 4\sin\dfrac{kz}{2}\sin\left(\dfrac{kx-ky+kz}{4}\right)\sin\left(\dfrac{kx-ky-kz}{4}\right)$

$= 4\sin\dfrac{kz}{2}\sin\left(\dfrac{-2ky}{4}\right)\sin\left(\dfrac{2kx}{4}\right)$

$= -4\sin\dfrac{kz}{2}\sin\dfrac{ky}{2}\sin\dfrac{kx}{2}.$

Therefoe, $\sin kx + \sin ky + \sin kz = -4\sin\dfrac{kx}{2}\sin\dfrac{ky}{2}\sin\dfrac{kz}{2}$.

2. $\cos kx + \cos ky + \cos kz = 4\cos\dfrac{kx}{2}\cos\dfrac{ky}{2}\cos\dfrac{kz}{2} - 1$;
Using addition and double angle formula, we have

$\cos kx + \cos ky + \cos kz$

$$= 2\cos\left(\frac{kx+ky}{2}\right)\cos\left(\frac{kx-ky}{2}\right) + 2\cos^2\frac{kz}{2} - 1$$

$$= 2\cos\left(\frac{-kz}{2}\right)\cos\left(\frac{kx-ky}{2}\right) + 2\cos^2\frac{kz}{2} - 1$$

$$= 2\cos\frac{kz}{2}\cos\left(\frac{kx-ky}{2}\right) + 2\cos^2\frac{kz}{2} - 1$$

$$= 2\cos\frac{kz}{2}\left[\cos\left(\frac{kx-ky}{2}\right) + \cos\frac{kz}{2}\right] - 1$$

$$= 2\cos\frac{kz}{2} \times 2\cos\left(\frac{\frac{kx-ky}{2} + \frac{kz}{2}}{2}\right)\cos\left(\frac{\frac{kx-ky}{2} - \frac{kz}{2}}{2}\right) - 1$$

$$= 4\cos\frac{kz}{2}\cos\left(\frac{kx-ky+kz}{4}\right)\cos\left(\frac{kx-ky-kz}{4}\right) - 1$$

$$= 4\cos\frac{kz}{2}\cos\left(\frac{-2ky}{4}\right)\cos\left(\frac{2kx}{4}\right) - 1$$

$$= 4\cos\frac{kz}{2}\cos\frac{ky}{2}\cos\frac{kx}{2} - 1.$$

Therefore,

$$\cos kx + \cos ky + \cos kz = 4\cos\frac{kx}{2}\cos\frac{ky}{2}\cos\frac{kz}{2} - 1.$$

3. $\tan kx + \tan ky + \tan kz = \tan kx \tan ky \tan kz$;
We have $kx + ky = -kz$. Then $\tan(kx+ky) = \tan(-kz)$.
Using addition formula, we have

$$\frac{\tan kx + \tan ky}{1 - \tan kx \tan ky} = -\tan kz.$$

Multiply both sides of the equality by $1 - \tan kx \tan ky$, we obtain

$$\tan kx + \tan ky = -\tan kz(1 - \tan kx \tan ky)$$

or

$$\tan kx + \tan ky = -\tan kz + \tan kx \tan ky \tan kz.$$

Therefore, $\tan kx + \tan ky + \tan kz = \tan kx \tan ky \tan kz$.

4. $\cot kx \cot ky + \cot ky \cot kz + \cot kz \cot kx = 1$
We have $kx + ky = -kz$. Then $\cot(kx + ky) = \cot(-kz)$.
Using addition formula, we have

$$\frac{\cot kx \cot ky - 1}{\cot kx + \cot ky} = -\cot kz.$$

Multiply both sides of the equality by $\cot kx + \cot ky$, we obtain

$$\cot kx \cot ky - 1 = -\cot kz (\cot kx + \cot ky)$$

or

$$\cot kx \cot ky - 1 = -\cot kz \cot kx - \cot kz \cot ky.$$

Therefore, $\cot kx \cot ky + \cot ky \cot kz + \cot kz \cot kx = 1$.

Remark 3. Let ABC be a triangle. From the above problem, we obtain

1. $\sin(A - B) + \sin(B - C) + \sin(C - A)$
$= -4 \sin \dfrac{A-B}{2} \sin \dfrac{B-C}{2} \sin \dfrac{C-A}{2}$;

2. $\cos(A - B) + \cos(B - C) + \cos(C - A)$
$= 4 \cos \dfrac{A-B}{2} \cos \dfrac{B-C}{2} \cos \dfrac{C-A}{2} - 1$;

3. $\tan(A - B) + \tan(B - C) + \tan(C - A)$
$= \tan(A - B) \tan(B - C) \tan(C - A)$;

4. $\cot(A - B) \cot(B - C) + \cot(B - C) \cot(C - A)$
$+ \cot(C - A) \cot(A - B) = 1$.

Problem 21. Let ABC be a triangle. Prove that

$$\sin^2 \frac{A}{2} \cos(B-C) + \sin^2 \frac{B}{2} \cos(C-A) + \sin^2 \frac{C}{2} \cos(A-B)$$
$$= 2 \cos \frac{A-B}{2} \cos \frac{B-C}{2} \cos \frac{C-A}{2} - 2 \cos A \cos B \cos C - 1.$$

Solution. Observe that

$$\sin^2 \frac{A}{2} \cos(B-C) + \sin^2 \frac{B}{2} \cos(C-A) + \sin^2 \frac{C}{2} \cos(A-B)$$

$$= \left(\frac{1-\cos A}{2}\right)\cos(B-C) + \left(\frac{1-\cos B}{2}\right)\cos(C-A)$$
$$+ \left(\frac{1-\cos C}{2}\right)\cos(A-B)$$
$$= \frac{1}{2}\cos(B-C) + \frac{1}{2}\cos(C-A) + \frac{1}{2}\cos(A-B)$$
$$- \frac{1}{2}\cos A\cos(B-C) - \frac{1}{2}\cos B\cos(C-A) - \frac{1}{2}\cos C\cos(A-B)$$
$$= \frac{1}{2}[\cos(A-B) + \cos(B-C) + \cos(C-A)]$$
$$- \frac{1}{2}[\cos A\cos(B-C) + \cos B\cos(C-A) + \cos C\cos(A-B)]. \tag{1}$$

Using product to sum formula, we have

$$\cos A\cos(B-C) = \frac{1}{2}[\cos(A-B+C) + \cos(A+B-C)]$$
$$= \frac{1}{2}[\cos(\pi - 2B) + \cos(\pi - 2C)]$$
$$= \frac{1}{2}(-\cos 2B - \cos 2C).$$

Similarly,

$$\cos B\cos(C-A) = \frac{1}{2}(-\cos 2C - \cos 2A)$$

and

$$\cos C\cos(A-B) = \frac{1}{2}(-\cos 2A - \cos 2B).$$

It follows that

$$\cos A\cos(B-C) + \cos B\cos(C-A) + \cos C\cos(A-B)$$
$$= \frac{1}{2}(-2\cos 2A - 2\cos 2B - 2\cos 2C)$$
$$= -(\cos 2A + \cos 2B + \cos 2C)$$
$$= -(-1 - 4\cos A\cos B\cos C)$$
$$= 1 + 4\cos A\cos B\cos C.$$

Furthermore,

$$\cos(A-B) + \cos(B-C) + \cos(C-A)$$

$$= 4\cos\frac{A-B}{2}\cos\frac{B-C}{2}\cos\frac{C-A}{2} - 1.$$

From (1), we obtain

$$\sin^2\frac{A}{2}\cos(B-C) + \sin^2\frac{B}{2}\cos(C-A) + \sin^2\frac{C}{2}\cos(A-B)$$
$$= \frac{1}{2}\left(4\cos\frac{A-B}{2}\cos\frac{B-C}{2}\cos\frac{C-A}{2} - 1\right)$$
$$- \frac{1}{2}(1 + 4\cos A\cos B\cos C)$$
$$= 2\cos\frac{A-B}{2}\cos\frac{B-C}{2}\cos\frac{C-A}{2} - \frac{1}{2} - \frac{1}{2} - 2\cos A\cos B\cos C$$
$$= 2\cos\frac{A-B}{2}\cos\frac{B-C}{2}\cos\frac{C-A}{2} - 2\cos A\cos B\cos C - 1.$$

Therefore,

$$\sin^2\frac{A}{2}\cos(B-C) + \sin^2\frac{B}{2}\cos(C-A) + \sin^2\frac{C}{2}\cos(A-B)$$
$$= 2\cos\frac{A-B}{2}\cos\frac{B-C}{2}\cos\frac{C-A}{2} - 2\cos A\cos B\cos C - 1.$$

Problem 22. Let A, B and C be the three angles of a triangle. Prove that:

1. $\sin A + \sin B + \sin C \leq \dfrac{3\sqrt{3}}{2}$;

2. $1 < \cos A + \cos B + \cos C \leq \dfrac{3}{2}$;

3. $\tan A + \tan B + \tan C \geq 3\sqrt{3}$ (A, B and C are acute angles);

4. $\cot A + \cot B + \cot C \geq \sqrt{3}$;

5. $\sin^2 A + \sin^2 B + \sin^2 C \leq \dfrac{9}{2}$;

6. $\cos^2 A + \cos^2 B + \cos^2 C \geq \dfrac{3}{4}$;

7. $\tan^2 A + \tan^2 B + \tan^2 C \geq 9$ (A, B and C are acute angles);

8. $1 < \sin\dfrac{A}{2} + \sin\dfrac{B}{2} + \sin\dfrac{C}{2} \leq \dfrac{3}{2}$;

Chapter 8. Solutions

9. $2 < \cos\dfrac{A}{2} + \cos\dfrac{B}{2} + \cos\dfrac{C}{2} \leq \dfrac{3\sqrt{3}}{2}$;

10. $\tan\dfrac{A}{2} + \tan\dfrac{B}{2} + \tan\dfrac{C}{2} \geq \sqrt{3}$;

11. $\dfrac{3}{4} \leq \sin^2\dfrac{A}{2} + \sin^2\dfrac{B}{2} + \sin^2\dfrac{C}{2} < 1$;

12. $2 < \cos^2\dfrac{A}{2} + \cos^2\dfrac{B}{2} + \cos^2\dfrac{C}{2} \leq \dfrac{9}{4}$;

13. $\tan^2\dfrac{A}{2} + \tan^2\dfrac{B}{2} + \tan^2\dfrac{C}{2} \geq 1$;

14. $\sin A \sin B \sin C \leq \dfrac{3\sqrt{3}}{8}$;

15. $\cos A \cos B \cos C \leq \dfrac{1}{8}$;

16. $\sin\dfrac{A}{2}\sin\dfrac{B}{2}\sin\dfrac{C}{2} \leq \dfrac{1}{8}$;

17. $\cos\dfrac{A}{2}\cos\dfrac{A}{2}\cos\dfrac{A}{2} \leq \dfrac{3\sqrt{3}}{8}$;

18. $\tan\dfrac{A}{2}\tan\dfrac{B}{2}\tan\dfrac{C}{2} \leq \dfrac{1}{3\sqrt{3}}$.

Solution. Prove that

1. $\sin A + \sin B + \sin C \leq \dfrac{3\sqrt{3}}{2}$

2. $1 < \cos A + \cos B + \cos C \leq \dfrac{3}{2}$
   By knowing that A, B and C are the three angles of a triangle, we obtain $\cos A + \cos B + \cos C = 1 + 4\sin\dfrac{A}{2}\sin\dfrac{B}{2}\sin\dfrac{C}{2}$.
   Additionally, $\sin\dfrac{A}{2}, \sin\dfrac{B}{2}$ and $\sin\dfrac{C}{2} > 0$. Then
   $$\cos A + \cos B + \cos C > 1. \qquad (1)$$
   Since $0 < \cos\left(\dfrac{A-B}{2}\right)$ and $\cos\left(\dfrac{C}{2} - \dfrac{\pi}{6}\right) \leq 1$, it follows that
   $$\cos A + \cos B + \cos C + \cos\dfrac{\pi}{3}$$

$$\leq 2\cos\left(\frac{A+B}{2}\right)\cos\left(\frac{A-B}{2}\right) + 2\cos\left(\frac{C}{2}+\frac{\pi}{6}\right)\cos\left(\frac{C}{2}-\frac{\pi}{6}\right)$$

$$\leq 2\cos\left(\frac{A+B}{2}\right) + 2\cos\left(\frac{C}{2}+\frac{\pi}{6}\right)$$

$$= 2\left[\cos\left(\frac{A+B}{2}\right) + \cos\left(\frac{C}{2}+\frac{\pi}{6}\right)\right].$$

We obtain

$$\cos A + \cos B + \cos C + \cos\frac{\pi}{3}$$

$$\leq 4\cos\left(\frac{A+B+C+\frac{\pi}{3}}{4}\right)\cos\left(\frac{A+B-C-\frac{\pi}{3}}{4}\right)$$

$$\leq 4\cos\left(\frac{\pi+\frac{\pi}{3}}{4}\right) = 4\cos\frac{\pi}{3}$$

because $0 < \cos\left(\dfrac{A+B-C-\frac{\pi}{3}}{4}\right) \leq 1$.

Then
$$\cos A + \cos B + \cos C \leq 3\cos\frac{\pi}{3} = \frac{3}{2}. \qquad (2)$$

From (1) and (2), it implies that

$$1 < \cos A + \cos B + \cos C \leq \frac{3}{2}.$$

3. $\tan A + \tan B + \tan C \geq 3\sqrt{3}$

   We know that A, B and C are acute angles. Then $\tan A, \tan B$ and $\tan C > 0$.

   Using AM-GM inequality, we obtain

   $$\tan A + \tan B + \tan C \geq 3\sqrt[3]{\tan A \tan B \tan C}.$$

   Then
   $$(\tan A + \tan B + \tan C)^3 \geq 27\left(\tan A \tan B \tan C\right)$$
   $$= 27\left(\tan A + \tan B + \tan C\right).$$

   It follows that $(\tan A + \tan B + \tan C)^2 \geq 27$.
   Consequently, $\tan A + \tan B + \tan C \geq 3\sqrt{3}$.

4. $\cot A + \cot B + \cot C \geq 3\sqrt{3}$
   Using the fact that
   $$(x+y+z)^2 \geq 3(xy+yz+zx)$$
   and
   $$\cot A \cot B + \cot B \cot C + \cot C \cot A = 1$$
   , it follows that
   $$(\cot A + \cot B + \cot C)^2$$
   $$\geq 3(\cot A \cot B + \cot B \cot C + \cot C \cot A)$$
   $$= 3.$$
   Consequently, $\cot A + \cot B + \cot C \geq 3\sqrt{3}$.

5. $\sin^2 A + \sin^2 B + \sin^2 C \leq \dfrac{9}{4}$
   We have
   $$\sin^2 A + \sin^2 B + \sin^2 C$$
   $$= \frac{1-\cos 2A}{2} + \frac{1-\cos 2B}{2} + 1 - \cos^2 C$$
   $$= 2 - \frac{\cos 2A + \cos 2B}{2} - \cos^2 C$$
   $$= 2 - \cos(A+B)\cos(A-B) - \cos^2 C$$
   $$= 2 - \cos^2 C + \cos C \cos(A-B)$$
   $$= 2 - \cos^2 C + \cos C \cos(A-B) - \frac{1}{4}\cos^2(A-B) + \frac{1}{4}\cos^2(A-B)$$
   $$= -\left[\cos C - \frac{1}{2}\cos(A-B)\right]^2 + \frac{1}{4}[1-\sin^2(A-B)] + 2$$
   $$= -\left[\cos C - \frac{1}{2}\cos(A-B)\right]^2 - \frac{1}{4}\sin^2(A-B) + \frac{9}{4} \leq \frac{9}{4}$$
   because
   $$-\left[\cos C - \frac{1}{2}\cos(A-B)\right]^2 \leq 0$$
   and
   $$-\sin^2(A-B) \leq 0.$$
   Consequently, $\sin^2 A + \sin^2 B + \sin^2 C \leq \dfrac{9}{4}$.

6. $\cos^2 A + \cos^2 B + \cos^2 C \geq \dfrac{3}{4}$

   We have $\sin^2 A + \sin^2 B + \sin^2 C \leq \dfrac{9}{4}$.
   Since $\sin^2 x + \cos^2 x = 1$, then $\sin^2 x = 1 - \cos^2 x$,
   It implies that $1 - \cos^2 A + 1 - \cos^2 B + 1 - \cos^2 C \leq \dfrac{9}{4}$.
   Then $\cos^2 A + \cos^2 B + \cos^2 C \geq 3 - \dfrac{9}{4} = \dfrac{3}{4}$.
   Therefore, $\cos^2 A + \cos^2 B + \cos^2 C \geq \dfrac{3}{4}$.

7. $\tan^2 A + \tan^2 B + \tan^2 C \geq 9$
   Using AM-GM inequality, we obtain
   $$\tan^2 A + \tan^2 B + \tan^2 C \geq 3\sqrt[3]{(\tan A \tan B \tan C)^2}.$$
   From 3: $\tan A \tan B \tan C \geq 3\sqrt{3}$, then
   $$\tan^2 A + \tan^2 B + \tan^2 C \geq 3\sqrt[3]{(3\sqrt{3})^2} = 9.$$
   Consequently, $\tan^2 A + \tan^2 B + \tan^2 C \geq 9$.

8. $1 < \sin \dfrac{A}{2} + \sin \dfrac{B}{2} + \sin \dfrac{C}{2} \leq \dfrac{3}{2}$
   We have $0 < \cos \dfrac{A}{2}, \cos \dfrac{B}{2} < 1$.
   It implies that
   $$\sin \dfrac{A}{2} + \sin \dfrac{B}{2} > \sin \dfrac{A}{2} \cos \dfrac{B}{2} + \sin \dfrac{B}{2} \cos \dfrac{A}{2}$$
   $$= \sin\left(\dfrac{A+B}{2}\right)$$
   $$= \cos \dfrac{C}{2}.$$
   We obtain
   $$\sin \dfrac{A}{2} + \sin \dfrac{B}{2} + \sin \dfrac{C}{2} > \sin \dfrac{C}{2} + \cos \dfrac{C}{2}$$
   $$= \sqrt{2} \sin\left(\dfrac{\pi}{4} + \dfrac{C}{2}\right)$$
   $$> \sqrt{2}\left(\dfrac{\sqrt{2}}{2}\right) = 1.$$

As a result,
$$1 < \sin\frac{A}{2} + \sin\frac{B}{2} + \sin\frac{C}{2}. \tag{1}$$
Moreover,
$$\sin\frac{A}{2} + \sin\frac{B}{2} + \sin\frac{C}{2} + \sin\frac{\pi}{6}$$
$$= 2\sin\left(\frac{A+B}{2}\right)\cos\left(\frac{A-B}{2}\right) + 2\sin\left(\frac{C}{4} + \frac{\pi}{12}\right)\cos\left(\frac{C}{4} - \frac{\pi}{12}\right)$$
$$\leq 2\sin\left(\frac{A+B}{4}\right) + 2\sin\left(\frac{C}{4} + \frac{\pi}{12}\right)$$
$$= 2\left[\sin\left(\frac{A+B}{4}\right) + \sin\left(\frac{C}{4} + \frac{\pi}{12}\right)\right]$$
$$= 4\sin\left(\frac{A+B+C+\frac{\pi}{3}}{8}\right)\cos\left(\frac{A+B-C-\frac{\pi}{3}}{8}\right)$$
$$\leq 4\sin\left(\frac{\pi + \frac{\pi}{3}}{8}\right)$$
$$= 4\sin\frac{\pi}{6} = 2$$

because $\cos\left(\frac{A-B}{4}\right)$, $\cos\left(\frac{C}{4} - \frac{\pi}{12}\right)$ and $\cos\left(\frac{A+B-C-\frac{\pi}{3}}{8}\right)$
take values in $[0, 1]$.
Hence,
$$\sin\frac{A}{2} + \sin\frac{B}{2} + \sin\frac{C}{2} \leq 2 - \frac{1}{2} = \frac{3}{2}. \tag{2}$$
From (1) and (2), it implies that
$$1 < \sin\frac{A}{2} + \sin\frac{B}{2} + \sin\frac{C}{2} \leq \frac{3}{2}.$$

9. $2 < \cos\frac{A}{2} + \cos\frac{B}{2} + \cos\frac{C}{2} \leq \frac{3\sqrt{3}}{2}.$
We have $0 < \cos\frac{A}{2} < 1$. Then $\cos\frac{A}{2} > \cos^2\frac{A}{2}$.
Similarly, $\cos\frac{B}{2} > \cos^2\frac{B}{2}$ and $\cos\frac{C}{2} > \cos^2\frac{C}{2}$.
Hence,
$$\cos\frac{A}{2} + \cos\frac{B}{2} + \cos\frac{C}{2} > \cos^2\frac{A}{2} + \cos^2\frac{B}{2} + \cos^2\frac{C}{2}$$

$$= \frac{1+\cos A}{2} + \frac{1+\cos B}{2} + \frac{1+\cos C}{2}$$
$$= \frac{3}{2} + \frac{1}{2}(\cos A + \cos B + \cos C)$$
$$> \frac{3}{2} + \frac{1}{2}$$
$$= 2$$

Then
$$\cos\frac{A}{2} + \cos\frac{B}{2} + \cos\frac{C}{2} > 2. \qquad (1)$$

The same as in 8, it follows that
$$\cos\frac{A}{2} + \cos\frac{B}{2} + \cos\frac{C}{2} \leq \frac{3\sqrt{3}}{2}. \qquad (2)$$

From (1) and (2), we obtain
$$2 < \cos\frac{A}{2} + \cos\frac{B}{2} + \cos\frac{C}{2} \leq \frac{3\sqrt{3}}{2}.$$

10. $\tan\frac{A}{2} + \tan\frac{B}{2} + \tan\frac{C}{2} \geq \sqrt{3}$

We have $x^2 + y^2 + z^2 \geq xy + yz + zx$.
Consequently,

$$\tan^2\frac{A}{2} + \tan^2\frac{B}{2} + \tan^2\frac{C}{2}$$
$$\geq \tan\frac{A}{2}\tan\frac{B}{2} + \tan\frac{B}{2}\tan\frac{C}{2} + \tan\frac{C}{2}\tan\frac{A}{2} = 1.$$

Moreover,
$$\left(\tan\frac{A}{2} + \tan\frac{B}{2} + \tan\frac{C}{2}\right)^2 = \tan^2\frac{A}{2} + \tan^2\frac{B}{2} + \tan^2\frac{C}{2} + 2.$$

It follows that $\left(\tan\frac{A}{2} + \tan\frac{B}{2} + \tan\frac{C}{2}\right)^2 \geq 1 + 2 = 3.$

Therefore, $\tan\frac{A}{2} + \tan\frac{B}{2} + \tan\frac{C}{2} \geq \sqrt{3}.$

11. $\frac{3}{4} \leq \sin^2\frac{A}{2} + \sin^2\frac{B}{2} + \sin^2\frac{C}{2} < 1$

We have
$$\sin^2\frac{A}{2} + \sin^2\frac{B}{2} + \sin^2\frac{C}{2} = \frac{1-\cos A}{2} + \frac{1-\cos B}{2} + \frac{1-\cos C}{2}$$

$$= \frac{3}{2} - \frac{1}{2}(\cos A + \cos B + \cos C).$$

By knowing that $1 < \cos A + \cos B + \cos C \le \frac{3}{2}$, we obtain

$$\frac{3}{2} - \frac{1}{2}\left(\frac{3}{2}\right) \le \sin^2 \frac{A}{2} + \sin^2 \frac{B}{2} + \sin^2 \frac{C}{2} < \frac{3}{2} - \frac{1}{2}.$$

Thus, $\frac{3}{4} \le \sin^2 \frac{A}{2} + \sin^2 \frac{B}{2} + \sin^2 \frac{C}{2} < 1.$

12. $2 < \cos^2 \frac{A}{2} + \cos^2 \frac{B}{2} + \cos^2 \frac{C}{2} \le \frac{9}{4}$

We have

$$\cos^2 \frac{A}{2} + \cos^2 \frac{B}{2} + \cos^2 \frac{C}{2} = 1 - \sin^2 \frac{A}{2} + 1 - \sin^2 \frac{B}{2} + 1 - \sin^2 \frac{C}{2}$$

$$= 3 - \left(\sin^2 \frac{A}{2} + \sin^2 \frac{B}{2} + \sin^2 \frac{C}{2}\right).$$

By knowing that $\frac{3}{4} \le \sin^2 \frac{A}{2} + \sin^2 \frac{B}{2} + \sin^2 \frac{A}{C} < 1$, we obtain

$$3 - 1 < \cos^2 \frac{A}{2} + \cos^2 \frac{B}{2} + \cos^2 \frac{C}{2} \le 3 - \frac{3}{4}.$$

Hence, $2 < \cos^2 \frac{A}{2} + \cos^2 \frac{B}{2} + \cos^2 \frac{C}{2} \le \frac{9}{4}.$

13. $\tan^2 \frac{A}{2} + \tan^2 \frac{B}{2} + \tan^2 \frac{C}{2} \ge 1$

Using Cauchy-Schwarz's inequality, we obtain

$$\left(\tan \frac{A}{2} + \tan \frac{B}{2} + \tan \frac{C}{2}\right)^2$$

$$\le (1^2 + 1^2 + 1^2)\left(\tan^2 \frac{A}{2} + \tan^2 \frac{B}{2} + \tan^2 \frac{C}{2}\right)$$

Then $\tan^2 \frac{A}{2} + \tan^2 \frac{B}{2} + \tan^2 \frac{C}{2} \ge \frac{1}{3}\left(\tan \frac{A}{2} + \tan \frac{B}{2} + \tan \frac{C}{2}\right)^2.$

By knowing that $\tan \frac{A}{2} + \tan \frac{B}{2} + \tan \frac{C}{2} \ge \sqrt{3}$, it follows that

$$\tan \frac{A}{2}^2 + \tan^2 \frac{B}{2} + \tan^2 \frac{C}{2} \ge \left(\frac{1}{3}\right)3 = 1.$$

Thus, the given inequality is proved.

14. $\sin A \sin B \sin C \leq \dfrac{3\sqrt{3}}{8}$

Using AM-GM inequality, we obtain

$$\sin A \sin B \sin C \leq \left(\dfrac{\sin A + \sin B + \sin C}{3}\right)^3.$$

Since $\sin A + \sin B + \sin C \leq \dfrac{3\sqrt{3}}{2}$,

$$\sin A \sin B \sin C \leq \left(\dfrac{\frac{3\sqrt{3}}{2}}{3}\right)^3 = \dfrac{3\sqrt{3}}{8}.$$

Therefore, $\sin A \sin B \sin C \leq \dfrac{3\sqrt{3}}{8}$.

15. $\cos A \cos B \cos C \leq \dfrac{1}{8}$

We have

$\cos A \cos B \cos C$
$= \dfrac{1}{2}[\cos(A+B) + \cos(A-B)]\cos C$
$= \dfrac{1}{2}[-\cos C + \cos(A-B)]\cos C$
$= -\dfrac{1}{2}\left[\cos C - \dfrac{1}{2}\cos(A-B)\right]^2 + \dfrac{1}{8}\cos^2(A-B)$
$= -\dfrac{1}{2}\left[\cos C - \dfrac{1}{2}\cos(A-B)\right]^2 - \dfrac{1}{8}\sin^2(A-B) + \dfrac{1}{8} \leq \dfrac{1}{8}.$

Consequently, $\cos A \cos B \cos C \leq \dfrac{1}{8}$.

16. $\sin \dfrac{A}{2} \sin \dfrac{B}{2} \sin \dfrac{C}{2} \leq \dfrac{1}{8}$

Using AM-GM inequality, we obtain

$$\sin \dfrac{A}{2} \sin \dfrac{B}{2} \sin \dfrac{C}{2} \leq \left(\dfrac{\sin \frac{A}{2} + \sin \frac{B}{2} + \sin \frac{C}{2}}{3}\right)^3.$$

Chapter 8. Solutions

By knowing that $\sin\frac{A}{2} + \sin\frac{B}{2} + \sin\frac{C}{2} \le \frac{3}{2}$, it implies that

$$\sin\frac{A}{2}\sin\frac{B}{2}\sin\frac{C}{2} \le \left(\frac{\frac{3}{2}}{3}\right)^3 = \frac{1}{8}.$$

Therefore, $\sin\frac{A}{2}\sin\frac{B}{2}\sin\frac{C}{2} \le \frac{1}{8}$.

17. $\cos\frac{A}{2}\cos\frac{B}{2}\cos\frac{C}{2} \le \frac{3\sqrt{3}}{8}$

Using AM-GM inequality,

$$\cos\frac{A}{2}\cos\frac{B}{2}\cos\frac{C}{2} \le \left(\frac{\cos\frac{A}{2} + \cos\frac{B}{2} + \cos\frac{C}{2}}{3}\right)^3.$$

Since $\cos\frac{A}{2} + \cos\frac{B}{2} + \cos\frac{C}{2} \le \frac{3\sqrt{3}}{2}$, then

$$\cos\frac{A}{2}\cos\frac{B}{2}\cos\frac{C}{2} \le \left(\frac{\frac{3\sqrt{3}}{2}}{3}\right)^3 = \frac{3\sqrt{3}}{8}.$$

Therefore, $\cos\frac{A}{2}\cos\frac{B}{2}\cos\frac{C}{2} \le \frac{3\sqrt{3}}{8}$.

18. $\tan\frac{A}{2}\tan\frac{B}{2}\tan\frac{C}{2} \le \frac{1}{3\sqrt{3}}$

Using AM-GM inequality,

$$\cot\frac{A}{2}\cot\frac{B}{2}\cot\frac{C}{2} \le \left(\frac{\cot\frac{A}{2} + \cot\frac{B}{2} + \cot\frac{C}{2}}{3}\right)^3.$$

Since $\cot\frac{A}{2} + \cot\frac{B}{2} + \cot\frac{C}{2} = \cot\frac{A}{2}\cot\frac{B}{2}\cot\frac{C}{2}$ (Problem 18), then

$$\cot\frac{A}{2}\cot\frac{B}{2}\cot\frac{C}{2} \le \frac{\left(\cot\frac{A}{2}\cot\frac{B}{2}\cot\frac{C}{2}\right)^3}{27}.$$

It follows that $\left(\dfrac{1}{\cot\dfrac{A}{2}\cot\dfrac{B}{2}\cot\dfrac{C}{2}}\right)^2 \leq \dfrac{1}{27}.$

Hence, $\tan\dfrac{A}{2}\tan\dfrac{B}{2}\tan\dfrac{C}{2} \leq \dfrac{1}{3\sqrt{3}}.$

Problem 23. Prove the following inequalities:

1. $\sin^4\theta + \cos^4\theta \geq \dfrac{1}{2}$;

2. $\sin^6\theta + \cos^6\theta \geq \dfrac{1}{4}$;

3. $\sin^8\theta + \cos^8\theta \geq \dfrac{1}{8}$;

4. $\sin^{2n}\theta + \cos^{2n}\theta \geq \dfrac{1}{2^{n-1}}$ for all positive integers n.

Solution. Prove the following inequalities:

1. $\sin^4\theta + \cos^4\theta \geq \dfrac{1}{2}$

We have
$$\sin^4\theta + \cos^4\theta = \left(\sin^2\theta + \cos^2\theta\right)^2 - 2\sin^2\theta\cos^2\theta$$
$$= 1 - \dfrac{1}{2}(2\sin\theta\cos\theta)^2$$
$$= 1 - \dfrac{1}{2}\sin^2 2\theta$$

From the fact that $0 \leq \sin^2 2\theta \leq 1$, we obtain
$$0 \geq -\dfrac{1}{2}\sin^2 2\theta \geq -\dfrac{1}{2}.$$

It follows that
$$\sin^4\theta + \cos^4\theta \geq 1 - \dfrac{1}{2} = \dfrac{1}{2}.$$

Thus, $\sin^4\theta + \cos^4\theta \geq \dfrac{1}{2}.$

2. $\sin^6\theta + \cos^6\theta \geq \dfrac{1}{4}$

Since $\sin^2\theta + \cos^2\theta = 1$ and $\sin^4\theta + \cos^4\theta \geq \dfrac{1}{2}$, we obtain

$$\left(\sin^2\theta + \cos^2\theta\right)\left(\sin^4\theta + \cos^4\theta\right) \geq \dfrac{1}{2}$$

or

$$\sin^6\theta + \sin^2\theta\cos^4\theta + \sin^4\theta\cos^2\theta + \cos^6\theta \geq \dfrac{1}{2}.$$

Then $\sin^6\theta + \cos^6\theta + \sin^2\theta\cos^2\theta\left(\sin^2\theta + \cos^2\theta\right) \geq \dfrac{1}{2}$.
It implies that

$$\sin^6\theta + \cos^6\theta \geq \dfrac{1}{2} - \sin^2\theta\cos^2\theta$$

$$= \dfrac{1}{2} - \dfrac{1}{4}\sin^2 2\theta$$

$$\geq \dfrac{1}{2} - \dfrac{1}{4} = \dfrac{1}{4}$$

Hence, the claim is proved.

3. $\sin^8\theta + \cos^8\theta \geq \dfrac{1}{8}$

We have $\sin^6\theta + \cos^6\theta \geq \dfrac{1}{4}$ and $\sin^2\theta + \cos^2\theta = 1$.
Then

$$\left(\sin^2\theta + \cos^2\theta\right)\left(\sin^6\theta + \cos^6\theta\right) \geq \dfrac{1}{4}$$

or

$$\sin^8\theta + \sin^2\theta\cos^6\theta + \sin^6\theta\cos^2\theta + \cos^8\theta \geq \dfrac{1}{4}.$$

We obtain

$$\sin^8\theta + \cos^8\theta \geq \dfrac{1}{4} - \sin^2\theta\cos^2\theta\left(\sin^4\theta + \cos^4\theta\right)$$

$$= \dfrac{1}{4} - \dfrac{1}{4}\sin^2 2\theta\left(\sin^4\theta + \cos^4\theta\right).$$

Since $\sin^4\theta + \cos^4\theta \geq \dfrac{1}{2}$ and $-\dfrac{1}{4}\sin^2 2\theta \geq -\dfrac{1}{4}$, it follows that

$$\sin^8\theta + \cos^8\theta \geq \dfrac{1}{4} - \dfrac{1}{4}\left(\dfrac{1}{2}\right) = \dfrac{1}{8}.$$

Therefore, $\sin^8\theta + \cos^8\theta \geq \frac{1}{8}$.

4. $\sin^{2n}\theta + \cos^{2n}\theta \geq \frac{1}{2^{n-1}}$ for all positive integers n.
   From the above proof, the given statement is true for $n = 2, 3$.
   Suppose that
   $$\sin^{2(n-1)}\theta + \cos^{2(n-1)}\theta \geq \frac{1}{2^{n-2}}$$
   and
   $$\sin^{2n}\theta + \cos^{2n}\theta \geq \frac{1}{2^{n-1}}.$$
   We shall show that $\sin^{2(n+1)}\theta + \cos^{2(n+1)}\theta \geq \frac{1}{2^n}$.
   Since $\sin^2\theta + \cos^2\theta = 1$ and $\sin^{2n}\theta + \cos^{2n}\theta \geq \frac{1}{2^{n-1}}$, it follows that
   $$\left(\sin^2\theta + \cos^2\theta\right)\left(\sin^{2n}\theta + \cos^{2n}\theta\right) \geq \frac{1}{2^{n-1}}$$
   $$\sin^{2n+2}\theta + \sin^2 x\cos^{2n}\theta + \sin^{2n}\theta\cos^2\theta + \cos^{2n+2}\theta \geq \frac{1}{2^{n-1}}$$
   $$\sin^{2(n+1)}\theta + \cos^{2(n+1)}\theta + \frac{1}{4}\sin^2 2\theta\left(\sin^{2n-2}\theta + \cos^{2n-2}\theta\right) \geq \frac{1}{2^{n-1}}$$
   $$\sin^{2(n+1)}\theta + \cos^{2(n+1)}\theta \geq \frac{1}{2^{n-1}} - \frac{1}{4}\sin^2 2\theta\left(\sin^{2n-2}\theta + \cos^{2n-2}\theta\right) \tag{1}$$

   We know that
   $$-\frac{1}{4}\sin^2 2\theta \geq -\frac{1}{4}$$
   and
   $$\sin^{2(n-1)}\theta + \cos^{2(n-1)}\theta \geq \frac{1}{2^{n-2}}.$$
   Then (1) implies that
   $$\sin^{2(n+1)}\theta + \cos^{2(n+1)}\theta \geq \frac{1}{2^{n-1}} - \frac{1}{4}\left(\frac{1}{2^{n-2}}\right)$$
   $$= \frac{2}{2^n} - \frac{1}{2^n} = \frac{1}{2^n}.$$

   Therefore, $\sin^{2n}\theta + \cos^{2n}\theta \geq \frac{1}{2^{n-1}}$ for all positive integers n.

Problem 24. Prove the following equalities:

1. $\tan^2 \dfrac{\pi}{12} + \tan^2 \dfrac{3\pi}{12} + \tan^2 \dfrac{5\pi}{12} = 15$;

2. $\sin^4 \dfrac{\pi}{16} + \sin^4 \dfrac{3\pi}{16} + \sin^4 \dfrac{5\pi}{16} + \sin^4 \dfrac{7\pi}{16} = \dfrac{3}{2}$.

Solution. Prove the following equalities:

1. $\tan^2 \dfrac{\pi}{12} + \tan^2 \dfrac{3\pi}{12} + \tan^2 \dfrac{5\pi}{12} = 15$

   Using half-angle formula, $\tan^2 \dfrac{x}{2} = \dfrac{1 - \cos x}{1 + \cos x}$.
   It follows that

$$\tan^2 \dfrac{\pi}{12} + \tan^2 \dfrac{3\pi}{12} + \tan^2 \dfrac{5\pi}{12} = \dfrac{1 - \cos \dfrac{\pi}{6}}{1 + \cos \dfrac{\pi}{6}} + \tan^2 \dfrac{\pi}{4} + \dfrac{1 - \cos \dfrac{5\pi}{6}}{1 + \cos \dfrac{5\pi}{6}}$$

$$= \dfrac{1 - \dfrac{\sqrt{3}}{2}}{1 + \dfrac{\sqrt{3}}{2}} + 1 + \dfrac{1 + \dfrac{\sqrt{3}}{2}}{1 - \dfrac{\sqrt{3}}{2}}$$

$$= \dfrac{2 - \sqrt{3}}{2 + \sqrt{3}} + \dfrac{2 + \sqrt{3}}{2 - \sqrt{3}} + 1$$

$$= \dfrac{\left(2 - \sqrt{3}\right)^2 + \left(2 + \sqrt{3}\right)^2}{\left(2 - \sqrt{3}\right)\left(2 + \sqrt{3}\right)} + 1$$

$$= \dfrac{4 - 4\sqrt{3} + 3 + 4 + 4\sqrt{3} + 3}{4 - 3} + 1$$

$$= 14 + 1 = 15.$$

   Consequently, $\tan^2 \dfrac{\pi}{12} + \tan^2 \dfrac{3\pi}{12} + \tan^2 \dfrac{5\pi}{12} = 15$.

2. $\sin^4 \dfrac{\pi}{16} + \sin^4 \dfrac{3\pi}{16} + \sin^4 \dfrac{5\pi}{16} + \sin^4 \dfrac{7\pi}{16} = \dfrac{3}{2}$

   Let $S = \sin^4 \dfrac{\pi}{16} + \sin^4 \dfrac{3\pi}{16} + \sin^4 \dfrac{5\pi}{16} + \sin^4 \dfrac{7\pi}{16}$.
   We have $\sin^2 x = \dfrac{1 - \cos 2x}{2}$. It follows that

$$\sin^4 x = \left(\dfrac{1 - \cos 2x}{2}\right)^2$$

$$= \frac{1 - 2\cos 2x + \cos^2 2x}{4}$$
$$= \frac{1}{4} - \frac{1}{2}\cos 2x + \frac{1}{4}\cos^2 2x$$
$$= \frac{1}{4} - \frac{1}{2}\cos 2x + \frac{1}{4}\left(\frac{1 + \cos 4x}{2}\right)$$
$$= \frac{1}{4} - \frac{1}{2}\cos 2x + \frac{1}{8} + \frac{1}{8}\cos 4x$$
$$= \frac{3}{8} - \frac{1}{2}\cos 2x + \frac{1}{8}\cos 4x.$$

Substitute x by $\frac{\pi}{16}, \frac{3\pi}{16}, \frac{5\pi}{16}$ and $\frac{7\pi}{16}$, we obtain

$$\sin^4 \frac{\pi}{16} = \frac{3}{8} - \frac{1}{2}\cos \frac{\pi}{8} + \frac{1}{8}\cos \frac{\pi}{4}$$
$$\sin^4 \frac{3\pi}{16} = \frac{3}{8} - \frac{1}{2}\cos \frac{3\pi}{8} + \frac{1}{8}\cos \frac{3\pi}{4}$$
$$\sin^4 \frac{5\pi}{16} = \frac{3}{8} - \frac{1}{2}\cos \frac{5\pi}{8} + \frac{1}{8}\cos \frac{5\pi}{4}$$
and $\sin^4 \frac{7\pi}{16} = \frac{3}{8} - \frac{1}{2}\cos \frac{7\pi}{8} + \frac{1}{8}\cos \frac{7\pi}{4}.$

Adding the last four equalities, it implies that

$$S = \frac{3}{2} - \frac{1}{2}\left(\cos \frac{\pi}{8} + \cos \frac{3\pi}{8} + \cos \frac{5\pi}{8} + \cos \frac{7\pi}{8}\right)$$
$$+ \frac{1}{8}\left(\cos \frac{\pi}{4} + \cos \frac{3\pi}{4} + \cos \frac{5\pi}{4} + \cos \frac{7\pi}{4}\right)$$
$$= \frac{3}{2} - \frac{1}{2}\left(\cos \frac{\pi}{8} + \cos \frac{3\pi}{8} - \cos \frac{3\pi}{8} - \cos \frac{\pi}{8}\right)$$
$$+ \frac{1}{8}\left(\cos \frac{\pi}{4} + \cos \frac{3\pi}{4} - \cos \frac{\pi}{4} - \cos \frac{3\pi}{4}\right)$$
$$= \frac{3}{2} - \frac{1}{2}(0) + \frac{1}{8}(0) = \frac{3}{2}.$$

Consequently, the claim is proved.

Problem 25. Simplify $\sin 4\theta - 4\sin 3\theta + 6\sin 2\theta - 4\sin \theta$.

Solution. We have

$$\sin 4\theta - 4\sin 3\theta + 6\sin 2\theta - 4\sin \theta$$

$$\begin{aligned}
&= \sin 4\theta + 6\sin 2\theta - 4\left(\sin 3\theta + \sin\theta\right)\\
&= 2\sin 2\theta\cos 2\theta + 6\sin 2\theta - 8\sin 2\theta\cos\theta\\
&= 2\left(\cos 2\theta + 3 - 4\cos\theta\right)\sin 2\theta\\
&= 2\left(2\cos^2\theta - 1 + 3 - 4\cos\theta\right)\sin 2\theta\\
&= 2\left(2\cos^2\theta - 4\cos\theta + 2\right)\sin 2\theta\\
&= 4\left(\cos^2\theta - 2\cos\theta + 1\right)\sin 2\theta\\
&= 4(1-\cos\theta)^2\sin 2\theta\\
&= 4\left(2\sin^2\frac{\theta}{2}\right)^2\sin 2\theta = 16\sin^4\frac{\theta}{2}\sin 2\theta.
\end{aligned}$$

Thus, $\sin 4\theta - 4\sin 3\theta + 6\sin 2\theta - 4\sin\theta = 16\sin^4\dfrac{\theta}{2}\sin 2\theta$.

Problem 26. Calculate the following expressions:

1. $\tan 9° - \tan 27° - \tan 63° + \tan 81°$;

2. $\dfrac{1}{2\sin 10°} - 2\sin 70°$;

3. $3\sin 15°\cos 15° + \dfrac{\sin 60°}{\sin^2 15° - \cos^2 15°}$.

Solution. Calculate the following expressions:

1. $\tan 9° - \tan 27° - \tan 63° + \tan 81°$

Using the formula, $\tan p + \tan q = \dfrac{\sin(p+q)}{\cos p\cos q}$, we obtain

$$\begin{aligned}
&\tan 9° - \tan 27° - \tan 63° + \tan 81°\\
&= (\tan 9° + \tan 81°) - (\tan 27° + \tan 61°)\\
&= \frac{\sin(9° + 81°)}{\cos 9°\cos 81°} - \frac{\sin(27° + 63°)}{\cos 27°\cos 63°}\\
&= \frac{\sin 90°}{\cos 9°\cos 81°} - \frac{\sin 90°}{\cos 27°\cos 63°}\\
&= \frac{1}{\cos 9°\cos 81°} - \frac{1}{\cos 27°\cos 63°}\\
&= \frac{2}{2\cos 9°\cos 81°} - \frac{2}{2\cos 27°\cos 63°}\\
&= \frac{2}{\cos(9°-81°) + \cos(9°+81°)}
\end{aligned}$$

$$-\frac{2}{\cos(27° - 63°) + \cos(27° + 63°)}$$
$$= \frac{2}{\cos 72° + \cos 90°} - \frac{2}{\cos 36° + \cos 90°}$$
$$= \frac{2}{\cos 72°} - \frac{2}{\cos 36°}$$
$$= 2\left(\frac{\cos 36° - \cos 72°}{\cos 36° \cos 72°}\right)$$
$$= 2\left[\frac{-2\sin\left(\frac{36° + 72°}{2}\right)\sin\left(\frac{36° - 72°}{2}\right)}{\cos 36° \cos 72°}\right]$$
$$= \frac{-4\sin 54° \sin(-18°)}{\cos 36° \cos 72°}$$
$$= \frac{4\cos 36° \cos 72°}{\cos 36° \cos 72°} = 4.$$

Consequently, $\tan 9° - \tan 27° - \tan 63° + \tan 81° = 4$.

2. $\dfrac{1}{2\sin 10°} - 2\sin 70°$
We have
$$\frac{1}{2\sin 10°} - 2\sin 70° = \frac{1 - 4\sin 10° \sin 70°}{2\sin 10°}$$
$$= \frac{1 - 2[\cos(10° - 70°) - \cos(10° + 70°)]}{2\sin 10°}$$
$$= \frac{1 - 2(\cos 60° - \cos 80°)}{2\sin 10°}$$
$$= \frac{1 - 2\left(\frac{1}{2} - \sin 10°\right)}{2\sin 10°}$$
$$= \frac{1 - 1 + 2\sin 10°}{2\sin 10°} = \frac{2\sin 10°}{2\sin 10°} = 1.$$

Therefore, $\dfrac{1}{2\sin 10°} - 2\sin 70° = 1$.

3. $3\sin 15° \cos 15° + \dfrac{\sin 60°}{\sin^2 15° - \cos^2 15°}$
We have
$$3\sin 15° \cos 15° + \frac{\sin 60°}{\sin^2 15° - \cos^2 15°}$$

$$= \frac{3}{2}(2\sin 15°\cos 15°) - \frac{\sin 60°}{\cos^2 15° - \sin^2 15°}$$
$$= \frac{3}{2}\sin 30° - \frac{\sin 60°}{\cos 30°}$$
$$= \frac{3}{2}\left(\frac{1}{2}\right) - \frac{\cos 30°}{\cos 30°}$$
$$= \frac{3}{4} - 1 = -\frac{1}{4}.$$

Thus, $3\sin 15°\cos 15° + \dfrac{\sin 60°}{\sin^2 15° - \cos^2 15°} = -\dfrac{1}{4}.$

Problem 27. Simplify the following expressions:

1. $\dfrac{1}{2}\tan\dfrac{\theta}{2} + \dfrac{1}{2^2}\tan\dfrac{\theta}{2^2} + ... + \dfrac{1}{2^n}\tan\dfrac{\theta}{2^n};$

2. $\dfrac{1}{4\cos^2\dfrac{\theta}{2}} + \dfrac{1}{4^2\cos^2\dfrac{\theta}{2^2}} + ... + \dfrac{1}{4^n\cos^2\dfrac{\theta}{2^n}};$

3. $\sin^3\dfrac{\theta}{3} + 3\sin^3\dfrac{\theta}{3^2} + 3^2\sin^3\dfrac{\theta}{3^3} + ... + 3^{n-1}\sin^3\dfrac{\theta}{3^n}.$

Solution. 1. Let $S = \dfrac{1}{2}\tan\dfrac{\theta}{2} + \dfrac{1}{2^2}\tan\dfrac{\theta}{2^2} + ... + \dfrac{1}{2^n}\tan\dfrac{\theta}{2^n}.$

Using double-angle formula, $\tan 2x = \dfrac{2\tan x}{1 - \tan^2 x}$, then

$$\cot 2x = \dfrac{1}{\tan 2x} = \dfrac{1 - \tan^2 x}{2\tan x}$$
$$= \dfrac{1}{2}\left(\dfrac{1}{\tan x} - \dfrac{\tan^2 x}{\tan x}\right)$$
$$= \dfrac{1}{2}(\cot x - \tan x)$$

or
$$2\cot 2x = \cot x - \tan x.$$

It follows that $\tan x = \cot x - 2\cot 2x.$
Substitute x by $\dfrac{\theta}{2}, \dfrac{\theta}{2^2}, \dfrac{\theta}{2^3}, ...,$ and $\dfrac{\theta}{2^n}$, we obtain

$$\dfrac{1}{2}\tan\dfrac{\theta}{2} = \dfrac{1}{2}\cot\dfrac{\theta}{2} - \cot\theta$$

$$\frac{1}{2^2}\tan\frac{\theta}{2^2} = \frac{1}{2^2}\cot\frac{\theta}{2^2} - \frac{1}{2}\cot\frac{\theta}{2}$$
$$\frac{1}{2^3}\tan\frac{\theta}{2^3} = \frac{1}{2^3}\cot\frac{\theta}{2^3} - \frac{1}{2^2}\cot\frac{\theta}{2^2}$$
$$\vdots$$
$$\frac{1}{2^n}\tan\frac{\theta}{2^n} = \frac{1}{2^n}\cot\frac{\theta}{2^n} - \frac{1}{2^{n-1}}\cot\frac{\theta}{2^{n-1}}.$$

Adding the equalities, it implies that $S = \frac{1}{2^n}\cot\frac{\theta}{2^n} - \cot\theta$.

2. Let $T = \dfrac{1}{4\cos^2\frac{\theta}{2}} + \dfrac{1}{4^2\cos^2\frac{\theta}{2^2}} + \dfrac{1}{4^3\cos^2\frac{\theta}{2^3}} + ... + \dfrac{1}{4^n\cos^2\frac{\theta}{2^n}}.$

Observe that

$$\frac{1}{\cos^2 x} + \frac{1}{\sin^2 x} = \frac{\sin^2 x + \cos^2 x}{\sin^2 x \cos^2 x}$$
$$= \frac{1}{\sin^2 x \cos^2 x}$$
$$= \frac{4}{(2\sin x \cos x)^2}$$
$$= \frac{4}{\sin^2 2x}.$$

Then $\dfrac{1}{\cos^2 x} = \dfrac{4}{\sin^2 2x} - \dfrac{1}{\sin^2 x}.$

Substitute x by $\dfrac{\theta}{2}, \dfrac{\theta}{2^2}, \dfrac{\theta}{2^3}, ...,$ and $\dfrac{\theta}{2^n}$, we obtain

$$\frac{1}{4\cos^2\frac{\theta}{2}} = \frac{1}{\sin^2\theta} - \frac{1}{4\sin^2\frac{\theta}{2}}$$
$$\frac{1}{4^2\cos^2\frac{\theta}{2^2}} = \frac{1}{4\sin^2\frac{\theta}{2}} - \frac{1}{4^2\sin^2\frac{\theta}{2^2}}$$
$$\frac{1}{4^3\cos^2\frac{\theta}{2^3}} = \frac{1}{4^2\sin^2\frac{\theta}{2^2}} - \frac{1}{4^3\sin^2\frac{\theta}{2^3}}$$
$$\vdots$$

and $\dfrac{1}{4^n\cos^2\dfrac{\theta}{2^n}} = \dfrac{1}{4^{n-1}\sin^2\dfrac{\theta}{2^{n-1}}} - \dfrac{1}{4^n\sin^2\dfrac{\theta}{2^n}}.$

Adding the equalities, it follows that $T = \dfrac{1}{\sin^2\theta} - \dfrac{1}{4^n\sin^2\dfrac{\theta}{2^n}}.$

3. Let $R = \sin^3\dfrac{\theta}{3} + 3\sin^3\dfrac{\theta}{3^2} + 3^2\sin^3\dfrac{\theta}{3^3} + \ldots + 3^{n-1}\sin^3\dfrac{\theta}{3^n}.$

We have $\sin 3x = 3\sin x - 4\sin^3 x.$

Then $4\sin^3 x = 3\sin x - \sin 3x$ or $\sin^3 x = \dfrac{3}{4}\sin x - \dfrac{1}{4}\sin 3x.$

Substitute x by $\dfrac{\theta}{3}, \dfrac{\theta}{3^2}, \dfrac{\theta}{3^3}, \ldots,$ and $\dfrac{\theta}{3^{n-1}},$ we obtain

$$\sin^3\dfrac{\theta}{3} = \dfrac{3}{4}\sin\dfrac{\theta}{3} - \dfrac{1}{4}\sin\theta$$

$$3\sin^3\dfrac{\theta}{3^2} = \dfrac{3^2}{4}\sin\dfrac{\theta}{3^2} - \dfrac{3}{4}\sin\dfrac{\theta}{3}$$

$$3^2\sin^3\dfrac{\theta}{3^3} = \dfrac{3^3}{4}\sin\dfrac{\theta}{3^3} - \dfrac{3^2}{4}\sin\dfrac{\theta}{3^2}$$

$$\vdots$$

and $\quad 3^{n-1}\sin^3\dfrac{\theta}{3^n} = \dfrac{3^n}{4}\sin\dfrac{\theta}{3^n} - \dfrac{3^{n-1}}{4}\sin\dfrac{\theta}{3^{n-1}}.$

Adding the equalities, we obtain $R = \dfrac{3^n}{4}\sin\dfrac{\theta}{3^n} - \dfrac{1}{4}\sin\theta.$

Problem 28. Prove that

$$\cos\dfrac{\pi}{15}\cos\dfrac{2\pi}{15}\cos\dfrac{3\pi}{15}\cos\dfrac{4\pi}{15}\cos\dfrac{5\pi}{15}\cos\dfrac{6\pi}{15}\cos\dfrac{7\pi}{15} = \dfrac{1}{2^7}.$$

Solution. Let $P = \cos\dfrac{\pi}{15}\cos\dfrac{2\pi}{15}\cos\dfrac{3\pi}{15}\cos\dfrac{4\pi}{15}\cos\dfrac{5\pi}{15}\cos\dfrac{6\pi}{15}\cos\dfrac{7\pi}{15}.$

Using the formula, $\sin 2x = 2\sin x \cos x,$ we obtain $\cos x = \dfrac{1}{2}\dfrac{\sin 2x}{\sin x}.$
Then

$$P = \left(\dfrac{1}{2}\dfrac{\sin\dfrac{2\pi}{15}}{\sin\dfrac{\pi}{15}}\right)\left(\dfrac{1}{2}\dfrac{\sin\dfrac{4\pi}{15}}{\sin\dfrac{2\pi}{15}}\right)\left(\dfrac{1}{2}\dfrac{\sin\dfrac{6\pi}{15}}{\sin\dfrac{3\pi}{15}}\right)\left(\dfrac{1}{2}\dfrac{\sin\dfrac{8\pi}{15}}{\sin\dfrac{4\pi}{15}}\right)\left(\dfrac{1}{2}\dfrac{\sin\dfrac{10\pi}{15}}{\sin\dfrac{5\pi}{15}}\right)$$

$$\times \left(\frac{1}{2} \frac{\sin \frac{12\pi}{15}}{\sin \frac{6\pi}{15}}\right) \left(\frac{1}{2} \frac{\sin \frac{14\pi}{15}}{\sin \frac{7\pi}{15}}\right)$$

$$= \frac{1}{2^7} \frac{\sin \frac{8\pi}{15} \sin \frac{10\pi}{15} \sin \frac{12\pi}{15} \sin \frac{14\pi}{15}}{\sin \frac{\pi}{15} \sin \frac{3\pi}{15} \sin \frac{5\pi}{15} \sin \frac{7\pi}{15}}.$$

By knowing that $\sin(\pi - x) = \sin x$, it implies that

$$P = \frac{1}{2^7} \frac{\sin \frac{7\pi}{15} \sin \frac{5\pi}{15} \sin \frac{3\pi}{15} \sin \frac{\pi}{15}}{\sin \frac{\pi}{15} \sin \frac{3\pi}{15} \sin \frac{5\pi}{15} \sin \frac{7\pi}{15}} = \frac{1}{2^7}.$$

Thus, the given equality is proved.

Problem 29. Simplify the following expressions:

1. $A = \cos \frac{\theta}{2} \cos \frac{\theta}{2^2} \cos \frac{\theta}{2^3} \ldots \cos \frac{\theta}{2^n}$;

2. $B = (2\cos\theta - 1)(2\cos 2\theta - 1)(2\cos 2^2\theta - 1)\ldots(2\cos 2^n\theta - 1)$;

3. $C = \left(1 + \frac{1}{\cos\theta}\right)\left(1 + \frac{1}{\cos 2\theta}\right)\left(1 + \frac{1}{\cos 2^2\theta}\right)\ldots\left(1 + \frac{1}{\cos 2^n\theta}\right).$

Solution. Simplify the following expressions:

1. $A = \cos \frac{\theta}{2} \cos \frac{\theta}{2^2} \cos \frac{\theta}{2^3} \ldots \cos \frac{\theta}{2^n}$
   Using double-angle formula, $\sin 2x = 2\sin x \cos x$, then

$$\cos x = \frac{1}{2} \frac{\sin 2x}{\sin x}.$$

Substitute x by $\frac{\theta}{2}, \frac{\theta}{2^2}, \frac{\theta}{2^3}, \ldots$, and $\frac{\theta}{2^n}$, we obtain

$$\cos \frac{\theta}{2} = \frac{1}{2} \frac{\sin \theta}{\sin \frac{\theta}{2}}$$

$$\cos \frac{\theta}{2^2} = \frac{1}{2} \frac{\sin \frac{\theta}{2}}{\sin \frac{\theta}{2^2}}$$

Chapter 8. Solutions

$$\cos\frac{\theta}{2^3} = \frac{1}{2}\frac{\sin\frac{\theta}{2^2}}{\sin\frac{\theta}{2^3}}$$

and $\cos\dfrac{\theta}{2^n} = \dfrac{1}{2}\dfrac{\sin\dfrac{\theta}{2^{n-1}}}{\sin\dfrac{\theta}{2^n}}$.

It follows that

$$A = \left(\frac{1}{2}\frac{\sin\theta}{\sin\frac{\theta}{2}}\right)\left(\frac{1}{2}\frac{\sin\frac{\theta}{2}}{\sin\frac{\theta}{2^2}}\right)\left(\frac{1}{2}\frac{\sin\frac{\theta}{2^2}}{\sin\frac{\theta}{2^3}}\right)\cdots\left(\frac{1}{2}\frac{\sin\frac{\theta}{2^{n-1}}}{\sin\frac{\theta}{2^n}}\right)$$

$$= \frac{1}{2^n}\frac{\sin\theta}{\sin\frac{\theta}{2^n}}.$$

Therefore, $A = \dfrac{1}{2^n}\dfrac{\sin\theta}{\sin\dfrac{\theta}{2^n}}$.

2. $B = (2\cos\theta - 1)(2\cos 2\theta - 1)(2\cos 2^2\theta - 1)\ldots(2\cos 2^n\theta - 1)$
Observe that

$$(2\cos x - 1)(2\cos x + 1) = 4\cos^2 x - 1$$
$$= 4\left(\frac{1+\cos 2x}{2}\right) - 1$$
$$= 2 + 2\cos 2x - 1$$
$$= 1 + 2\cos 2x.$$

Then $2\cos x - 1 = \dfrac{2\cos 2x + 1}{2\cos x + 1}$.

Substitute x by $\theta, 2\theta, 2^2\theta, \ldots$, and $2^n\theta$, we obtain

$$2\cos\theta - 1 = \frac{2\cos 2\theta + 1}{2\cos\theta + 1}$$
$$2\cos 2\theta - 1 = \frac{2\cos 2^2\theta + 1}{2\cos 2\theta + 1}$$
$$2\cos 2^2\theta - 1 = \frac{2\cos 2^3\theta + 1}{2\cos 2^2\theta + 1}$$

and $\quad 2\cos 2^n\theta - 1 = \dfrac{2\cos 2^{n+1}\theta + 1}{2\cos 2^n\theta + 1}.$

It implies that

$$B = \left(\dfrac{2\cos 2\theta + 1}{2\cos\theta + 1}\right)\left(\dfrac{2\cos 2^2\theta + 1}{2\cos 2\theta + 1}\right)\left(\dfrac{2\cos 2^3\theta + 1}{2\cos 2^2\theta + 1}\right) \times \ldots$$
$$\times \left(\dfrac{2\cos 2^{n+1}\theta + 1}{2\cos 2^n\theta + 1}\right)$$
$$= \dfrac{2\cos 2^{n+1}\theta + 1}{2\cos\theta + 1}.$$

Therefore, $B = \dfrac{2\cos 2^{n+1}\theta + 1}{2\cos\theta + 1}.$

3. $C = \left(1 + \dfrac{1}{\cos\theta}\right)\left(1 + \dfrac{1}{\cos 2\theta}\right)\left(1 + \dfrac{1}{\cos 2^2\theta}\right)\ldots\left(1 + \dfrac{1}{\cos 2^n\theta}\right)$

Observe that

$$1 + \dfrac{1}{\cos x} = \dfrac{1 + \cos x}{\cos x}$$
$$= \dfrac{2\cos^2\dfrac{x}{2}}{\cos x}$$
$$= \dfrac{2\cos\dfrac{x}{2}\sin\dfrac{x}{2}\cos\dfrac{x}{2}}{\sin\dfrac{x}{2}\cos x}$$
$$= \dfrac{\cos\dfrac{x}{2}}{\sin\dfrac{x}{2}} \times \dfrac{\sin x}{\cos x} = \dfrac{\tan x}{\tan\dfrac{x}{2}}.$$

Substitute x by $\theta, 2\theta, 2^2\theta, \ldots$, and $2^n\theta$, we obtain

$$1 + \dfrac{1}{\cos\theta} = \dfrac{\tan\theta}{\tan\dfrac{\theta}{2}}$$
$$1 + \dfrac{1}{\cos 2\theta} = \dfrac{\tan 2\theta}{\tan\theta}$$
$$1 + \dfrac{1}{\cos 2^2\theta} = \dfrac{\tan 2^2\theta}{\tan 2\theta}$$
$$\text{and}\quad 1 + \dfrac{1}{\cos 2^n\theta} = \dfrac{\tan 2^n\theta}{\tan 2^{n-1}\theta}.$$

It follows that

$$C = \frac{\tan \theta}{\tan \frac{\theta}{2}} \times \frac{\tan 2\theta}{\tan \theta} \times \frac{\tan 2^2 \theta}{\tan 2\theta} \times \ldots \times \frac{\tan 2^n \theta}{\tan 2^{n-1} \theta} = \frac{\tan 2^n \theta}{\tan \frac{\theta}{2}}.$$

Therefore, $C = \dfrac{\tan 2^n \theta}{\tan \frac{\theta}{2}}$.

Problem 30. Given a triangle ABC. Prove that $\sin \dfrac{A}{2} \leq \dfrac{a}{b+c}$, $\sin \dfrac{B}{2} \leq \dfrac{b}{c+a}$ and $\sin \dfrac{C}{2} \leq \dfrac{a+b}{2}$.

Solution. To prove the statement, it is sufficient to prove that $\sin \dfrac{A}{2} \leq \dfrac{a}{b+c}$, namely, using the same proof for the rest.

Using the law of sine, $\dfrac{a}{\sin A} = \dfrac{b}{\sin B} = \dfrac{c}{\sin C} = 2R$, then $a = 2R \sin A$, $b = 2R \sin B$ and $c = 2R \sin C$.
Consequently,

$$\frac{a}{b+c} = \frac{2R \sin A}{2R \sin B + 2R \sin C}$$
$$= \frac{2R \sin A}{2R (\sin B + \sin C)}$$
$$= \frac{\sin A}{\sin B + \sin C}$$
$$= \frac{2 \sin \dfrac{A}{2} \cos \dfrac{A}{2}}{2 \sin \left(\dfrac{B+C}{2} \right) \cos \left(\dfrac{B-C}{2} \right)}$$
$$= \frac{\sin \dfrac{A}{2} \cos \dfrac{A}{2}}{\cos \dfrac{A}{2} \cos \left(\dfrac{B-C}{2} \right)}$$
$$= \frac{\sin \dfrac{A}{2}}{\cos \left(\dfrac{B-C}{2} \right)}.$$

Since $0 < \cos\left(\dfrac{B-C}{2}\right) \leq 1$, we obtain $\dfrac{a}{b+c} \geq \sin\dfrac{A}{2}$.

Therefore, $\sin\dfrac{A}{2} \leq \dfrac{a}{b+c}$.

Problem 31. Show that $\dfrac{\sin(a-b)}{\cos a \cos b} + \dfrac{\sin(b-c)}{\cos b \cos c} + \dfrac{\sin(c-a)}{\cos c \cos a} = 0$.

Solution. We have
$$\dfrac{\sin(a-b)}{\cos a \cos b} + \dfrac{\sin(b-c)}{\cos b \cos c} + \dfrac{\sin(c-a)}{\cos c \cos a}$$
$$= \tan a - \tan b + \tan b - \tan c + \tan c - \tan a = 0$$

Consequently, $\dfrac{\sin(a-b)}{\cos a \cos b} + \dfrac{\sin(b-c)}{\cos b \cos c} + \dfrac{\sin(c-a)}{\cos c \cos a} = 0$.

Problem 32. Prove that $\cos\dfrac{\pi}{7} + \cos\dfrac{3\pi}{7} + \cos\dfrac{5\pi}{7} = \dfrac{1}{2}$.

Solution. Let $S = \cos\dfrac{\pi}{7} + \cos\dfrac{5\pi}{7} + \cos\dfrac{3\pi}{7}$.

It implies that $2S\sin\dfrac{\pi}{7} = 2\sin\dfrac{\pi}{7}\cos\dfrac{\pi}{7} + 2\sin\dfrac{\pi}{7}\cos\dfrac{3\pi}{7} + 2\sin\dfrac{\pi}{7}\cos\dfrac{5\pi}{7}$.

Moreover, $2\sin a \cos b = \sin(a+b) + \sin(a-b) = \sin(b+a) - \sin(b-a)$.

It follows that

$2\sin\dfrac{\pi}{7}\cos\dfrac{\pi}{7} = \sin\left(\dfrac{\pi}{7}+\dfrac{\pi}{7}\right) - \sin\left(\dfrac{\pi}{7}-\dfrac{\pi}{7}\right) = \sin\dfrac{2\pi}{7} - \sin 0$

$2\sin\dfrac{\pi}{7}\cos\dfrac{3\pi}{7} = \sin\left(\dfrac{3\pi}{7}+\dfrac{\pi}{7}\right) - \sin\left(\dfrac{3\pi}{7}-\dfrac{\pi}{7}\right) = \sin\dfrac{4\pi}{7} - \sin\dfrac{2\pi}{7}$

$2\sin\dfrac{\pi}{7}\cos\dfrac{5\pi}{7} = \sin\left(\dfrac{5\pi}{7}+\dfrac{\pi}{7}\right) - \sin\left(\dfrac{5\pi}{7}-\dfrac{\pi}{7}\right) = \sin\dfrac{6\pi}{7} - \sin\dfrac{4\pi}{7}$.

Adding the equalities, we obtain
$$2S\sin\dfrac{\pi}{7} = \sin\dfrac{6\pi}{7} - \sin 0 = \sin\left(\pi - \dfrac{\pi}{7}\right) = \sin\dfrac{\pi}{7}.$$

Then $S = \dfrac{1}{2}$.

Thus, the given equality is proved.

Problem 33. Simplify the following expression:
$$S = \dfrac{1}{\cos a \cos(a+x)} + \dfrac{1}{\cos(a+x)\cos(a+2x)} + \ldots$$
$$+ \dfrac{1}{\cos[a+(n-1)x]\cos(a+nx)}.$$

Chapter 8. Solutions

Solution. Observe that
$$\frac{1}{\cos\left[a+(k-1)x\right]\cos(a+kx)}$$
$$=\frac{1}{\sin x}\times\frac{\sin\left[a+kx-a-(k-1)x\right]}{\cos\left[a+(k-1)x\right]\cos(a+kx)}$$
$$=\frac{1}{\sin x}\left[\tan(a+kx)-\tan(a+(k-1)x)\right].$$

Substitute k by 1, 2, 3, ..., and n, we obtain

$$\frac{1}{\cos a \cos(a+x)}=\frac{1}{\sin x}\left[\tan(a+x)-\tan a\right]$$

$$\frac{1}{\cos(a+x)\cos(a+2x)}=\frac{1}{\sin x}\left[\tan(a+2x)-\tan(a+x)\right]$$

$$\frac{1}{\cos(a+2x)\cos(a+3x)}=\frac{1}{\sin x}\left[\tan(a+3x)-\tan(a+2x)\right]$$

$\vdots$

and $\dfrac{1}{\cos\left[a+(n-1)x\right]\cos(a+nx)}$
$$=\frac{1}{\sin x}\left[\tan(a+nx)-\tan(a+(n-1)x)\right].$$

Adding the equalities, it follows that $S=\dfrac{\tan(a+nx)-\tan a}{\sin x}$.

Therefore, $S=\dfrac{\tan(a+nx)-\tan a}{\sin x}$.

Remark 4. Taking $x=a$, we obtain
$$\frac{1}{\cos a \cos 2a}+\frac{1}{\cos 2a \cos 3a}+...+\frac{1}{\cos na \cos(n+1)a}$$
$$=\frac{\tan(n+1)a-\tan a}{\sin a}.$$

Problem 34. Prove that $\sin a+\sin(a+x)+\sin(a+2x)+...+\sin\left[a+(n-1)x\right]=\dfrac{\sin\dfrac{nx}{2}\cos\left[a+\left(\dfrac{n-1}{2}\right)x\right]}{\sin\dfrac{x}{2}}.$

Solution. Let $S=\sin a+\sin(a+x)+\sin(a+2x)+...+\sin\left[a+(n-1)x\right]$. Then
$$2S\sin\frac{x}{2}=2\sin\frac{x}{2}\sin a+2\sin\frac{x}{2}\sin(a+x)+...$$

$$+ 2\sin\frac{x}{2}\sin\left[a + (n-1)x\right].$$

Observe that

$$2\sin a \sin b = \cos(a-b) - \cos(a+b)$$

or

$$2\sin a \sin b = \cos(b-a) - \cos(b+a).$$

It follows that

$$2\sin\frac{x}{2}\sin[a+(k-1)x]$$
$$= \cos\left[a+(k-1)x-\frac{x}{2}\right] - \cos\left[a+(k-1)x+\frac{x}{2}\right]$$
$$= \cos\left[a+\left(k-\frac{3}{2}\right)x\right] - \cos\left[a+\left(k-\frac{1}{2}\right)x\right].$$

Substitute k by $1, 2, 3, \ldots,$ and n, it implies that

$$2\sin\frac{x}{2}\sin a = \cos\left(a-\frac{x}{2}\right) - \cos\left(a+\frac{x}{2}\right)$$
$$2\sin\frac{x}{2}\sin(a+x) = \cos\left(a+\frac{x}{2}\right) - \cos\left(a+\frac{3x}{2}\right)$$
$$2\sin\frac{x}{2}\sin(a+2x) = \cos\left(a+\frac{3x}{2}\right) - \cos\left(a+\frac{5x}{2}\right)$$
$$\vdots$$

and $2\sin\frac{x}{2}\sin[a+(n-1)x]$

$$= \cos\left[a+\left(n-\frac{3}{2}\right)x\right] - \cos\left[a+\left(n-\frac{1}{2}\right)x\right].$$

Adding the equalities, we obtain

$$2S\sin\frac{x}{2} = \cos\left(a-\frac{x}{2}\right) - \cos\left[a+\left(n-\frac{1}{2}\right)x\right]$$
$$= -2\sin\left[\frac{a-\frac{x}{2}-a-\left(n-\frac{1}{2}\right)x}{2}\right]$$

$$\times \sin\left[\frac{a - \frac{x}{2} + a + \left(n - \frac{1}{2}\right)x}{2}\right]$$

$$= -2\sin\left(\frac{-nx}{2}\right)\sin\left[\frac{2a + (n-1)x}{2}\right]$$

$$= 2\sin\frac{nx}{2}\sin\left[a + \left(\frac{n-1}{2}\right)x\right].$$

Then $S = \dfrac{\sin\dfrac{nx}{2}\sin\left[a + \left(\dfrac{n-1}{2}\right)x\right]}{\sin\dfrac{x}{2}}$.

Therefore, $\sin a + \sin(a + x) + \sin(a + 2x) + \ldots + \sin[a + (n-1)x]$

$$= \frac{\sin\dfrac{nx}{2}\sin\left[a + \left(\dfrac{n-1}{2}\right)x\right]}{\sin\dfrac{x}{2}}.$$

Problem 35. Prove that $\cos a + \cos(a + x) + \ldots + \cos[a + (n-1)x]$

$$= \frac{\sin\dfrac{nx}{2}\cos\left[a + \left(\dfrac{n-1}{2}\right)x\right]}{\sin\dfrac{x}{2}}.$$

Solution. Let $T = \cos a + \cos(a + x) + \ldots + \cos[a + (n-1)x]$. Then

$$2T\sin\frac{x}{2} = 2\sin\frac{x}{2}\cos a + 2\sin\frac{x}{2}\cos(a + x) + \ldots$$
$$+ 2\sin\frac{x}{2}\cos[a + (n-1)x].$$

By $2\sin a \cos b = \sin(b + a) - \sin(b - a)$, then

$$2\sin\frac{x}{2}\cos[a + (k-1)x]$$
$$= \sin\left[a + (k-1)x + \frac{x}{2}\right] - \sin\left[a + (k-1)x - \frac{x}{2}\right]$$
$$= \sin\left[a + \left(k - \frac{1}{2}\right)x\right] - \sin\left[a + \left(k - \frac{3}{2}\right)x\right].$$

Substitute k by 1, 2, 3, ..., and n, it follows that

$$2\sin\frac{x}{2}\cos a = \sin\left(a + \frac{x}{2}\right) - \sin\left(a - \frac{x}{2}\right)$$

$$2\sin\frac{x}{2}\cos(a+x) = \sin\left(a+\frac{3x}{2}\right) - \sin\left(a+\frac{x}{2}\right)$$

$$2\sin\frac{x}{2}\cos(a+2x) = \sin\left(a+\frac{5x}{2}\right) - \sin\left(a+\frac{3x}{2}\right)$$

$$\vdots$$

and $\quad 2\sin\dfrac{x}{2}\cos[a+(n-1)x]$

$$= \sin\left[a+\left(n-\frac{1}{2}\right)x\right] - \sin\left[a+\left(n-\frac{3}{2}\right)x\right].$$

Adding the equalities, we obtain

$$2T\sin\frac{x}{2} = \sin\left[a+\left(n-\frac{1}{2}\right)x\right] - \sin\left(a-\frac{x}{2}\right)$$

$$= 2\sin\left[\frac{a+\left(n-\frac{1}{2}\right)x - a + \frac{x}{2}}{2}\right]$$

$$\times \cos\left[\frac{a+\left(n-\frac{1}{2}\right)x + a - \frac{x}{2}}{2}\right]$$

$$= 2\sin\frac{nx}{2}\cos\left[\frac{2a+(n-1)x}{2}\right]$$

$$= 2\sin\frac{nx}{2}\cos\left[a+\left(\frac{n-1}{2}\right)x\right].$$

Hence, $T = \dfrac{\sin\dfrac{nx}{2}\cos\left[a+\left(\dfrac{n-1}{2}\right)x\right]}{\sin\dfrac{x}{2}}.$

Problem 36. Show that $\cos\dfrac{2\pi}{9}, \cos\dfrac{4\pi}{9}, \cos\dfrac{6\pi}{9}$ and $\cos\dfrac{8\pi}{9}$ are the roots of

$$16x^4 + 8x^3 - 12x^2 - 4x + 1 = 0.$$

Evaluate
$$\cos\frac{2\pi}{9} + \cos\frac{4\pi}{9} + \cos\frac{6\pi}{9} + \cos\frac{8\pi}{9}$$

and
$$\cos\frac{2\pi}{9}\cos\frac{4\pi}{9}\cos\frac{6\pi}{9}\cos\frac{8\pi}{9}.$$

Solution. Note that $\cos\frac{2n\pi}{9}$ has only five different values for all integers n.
They are $1, \cos\frac{2\pi}{9}, \cos\frac{4\pi}{9}, \cos\frac{6\pi}{9}$ and $\cos\frac{8\pi}{9}$.
Let $\theta = \frac{2n\pi}{9}$. Then $9\theta = 2n\pi$ or $5\theta = 2n\pi - 4\theta$. We obtain

$$\cos 5\theta = \cos(2n\pi - 4\theta)$$
$$\cos 5\theta + \cos\theta = \cos 4\theta + \cos\theta$$
$$2\cos 3\theta \cos 2\theta = 2\cos^2 2\theta - 1 + \cos\theta.$$

It follows that $2\left(4\cos^3\theta - 3\cos\theta\right)\left(2\cos^2\theta - 1\right)$.
$$= 2\left(2\cos^2\theta - 1\right)^2 - 1 + \cos\theta.$$
It implies that

$2\left(8\cos^5\theta - 4\cos^3\theta - 6\cos^3\theta + 3\cos\theta\right) = 2\left(4\cos^4\theta - 4\cos^2\theta + 1\right) - 1$
$+ \cos\theta$
$2\left(8\cos^5\theta - 10\cos^3\theta + 3\cos\theta\right) = 8\cos^4\theta - 8\cos^2\theta + \cos\theta + 1$
$16\cos^5\theta - 20\cos^3\theta + 6\cos\theta = 8\cos^4\theta - 8\cos^2\theta + \cos\theta + 1$
$16\cos^5\theta - 8\cos^4\theta - 20\cos^3\theta + 8\cos^2\theta + 5\cos\theta - 1 = 0.$

It follows that $16x^5 - 8x^4 - 20x^3 + 8x^2 + 5x - 1 = 0$, where $x = \cos\theta$.
The roots of this equation are $1, \cos\frac{2\pi}{9}, \cos\frac{4\pi}{9}, \cos\frac{6\pi}{9}$ and $\cos\frac{8\pi}{9}$.
Moreover, $16x^5 - 8x^4 - 20x^3 + 8x^2 + 5x - 1 = 0$ is equivalent to
$$(x-1)\left(16x^4 + 8x^3 - 12x^2 - 4x + 1\right) = 0.$$
Since $x = 1$ is the root of $x - 1$, we obtain $\cos\frac{2\pi}{9}, \cos\frac{4\pi}{9}, \cos\frac{6\pi}{9}$
and $\cos\frac{8\pi}{9}$ are the roots of $16x^4 + 8x^3 - 12x^2 - 4x + 1 = 0$.
From Vieta's theorem, we obtain
$$\cos\frac{2\pi}{9} + \cos\frac{4\pi}{9} + \cos\frac{6\pi}{9} + \cos\frac{8\pi}{9} = -\frac{8}{16} = -\frac{1}{2}$$
and
$$\cos\frac{2\pi}{9}\cos\frac{4\pi}{9}\cos\frac{6\pi}{9}\cos\frac{8\pi}{9} = \frac{1}{6}.$$

154

Problem 37. Solve the following equations:
1. $\sin 3x + \sin x = 0$;
2. $\sin 4x + \sin 2x = 0$;
3. $\cos 3x + \cos 2x = 0$;
4. $\cos 5x + \cos 3x = 0$;
5. $\tan 4x + \tan x = 0$.

Solution. Solve the equations:
1. $\sin 3x + \sin x = 0$
   Using sum to product formula,
   $$\sin p + \sin q = 2 \sin\left(\frac{p+q}{2}\right) \cos\left(\frac{p-q}{2}\right)$$
   , we obtain $\sin 3x + \sin x = 0$ is equivalent to
   $$2 \sin\left(\frac{3x+x}{2}\right) \cos\left(\frac{3x-x}{2}\right) = 0.$$
   Then $\sin 2x \cos x = 0$.
   It follows that $\left[\begin{array}{l} \sin 2x = 0 \\ \cos x = 0 \end{array}\right.$ or $\left[\begin{array}{l} 2x = k\pi \\ x = \frac{\pi}{2} + k\pi \end{array}\right.$.
   Consequently, $x = \frac{k\pi}{2}$ or $x = \frac{\pi}{2} + k\pi$.
   Hence, $x = \frac{k\pi}{2}$, where $k \in \mathbb{Z}$.

2. $\sin 4x + \sin 2x = 0$
   Using sum to product formula, we obtain
   $$2 \sin\left(\frac{4x+2x}{2}\right) \cos\left(\frac{4x-2x}{2}\right) = 0$$
   or
   $\sin 3x \cos x = 0.$
   It turns out that $\left[\begin{array}{l} \sin 3x = 0 \\ \cos x = 0 \end{array}\right.$ or $\left[\begin{array}{l} 3x = k\pi \\ x = \frac{\pi}{2} + k\pi \end{array}\right.$.
   Hence, $\left[\begin{array}{l} x = \frac{k\pi}{3} \\ x = \frac{\pi}{2} + k\pi \end{array}\right.$.
   Therefore, $x = \frac{k\pi}{3}$ or $x = \frac{\pi}{2} + k\pi$, where $k \in \mathbb{Z}$.

3. $\cos 3x + \cos 2x = 0$

Using the formula, $\cos p + \cos q = 2\cos\left(\dfrac{p+q}{2}\right)\cos\left(\dfrac{p-q}{2}\right)$, from $\cos 3x + \cos 2x = 0$, it follows that

$2\cos\left(\dfrac{3x+2x}{2}\right)\cos\left(\dfrac{3x-2x}{2}\right) = 0$ or $\cos\dfrac{5x}{2}\cos\dfrac{x}{2} = 0$.

It implies that $\left[\begin{array}{l}\cos\dfrac{5x}{2} = 0 \\ \cos\dfrac{x}{2} = 0\end{array}\right.$ or $\left[\begin{array}{l}\dfrac{5x}{2} = \dfrac{\pi}{2} + k\pi \\ \dfrac{x}{2} = \dfrac{\pi}{2} + k\pi\end{array}\right.$.

Consequently, $x = \dfrac{\pi}{5} + \dfrac{2k\pi}{5}$ or $x = \pi + 2k\pi$, where $k \in \mathbb{Z}$.

4. $\cos 5x + \cos 3x = 0$

Using sum to product formula, we obtain

$$2\cos\left(\dfrac{5x+3x}{2}\right)\cos\left(\dfrac{5x-3x}{2}\right) = 0$$

or
$$\cos 4x \cos x = 0.$$

It implies that $\left[\begin{array}{l}\cos 4x = 0 \\ \cos x = 0\end{array}\right.$. Then $\left[\begin{array}{l}4x = \dfrac{\pi}{2} + k\pi \\ x = \dfrac{\pi}{2} + k\pi\end{array}\right.$.

It follows that $\left[\begin{array}{l}x = \dfrac{\pi}{8} + \dfrac{k\pi}{4} \\ x = \dfrac{\pi}{2} + k\pi\end{array}\right.$.

Therefore, $x = \dfrac{\pi}{8} + \dfrac{k\pi}{4}$ or $x = \dfrac{\pi}{2} + k\pi$, where $k \in \mathbb{Z}$.

5. $\tan 4x + \tan x = 0$

Notice that $\tan 4x + \tan x = 0$ is defined if and only if

$\begin{cases}\cos 4x \neq 0 \\ \cos x \neq 0\end{cases}$ or $\begin{cases}4x \neq \dfrac{\pi}{2} + k\pi \\ x \neq \dfrac{\pi}{2} + k\pi\end{cases}$.

Then $\begin{cases} x \neq \dfrac{\pi}{8} + \dfrac{k\pi}{4} \\ x \neq \dfrac{\pi}{2} + k\pi \end{cases}$.

Moreover, $\tan 4x + \tan x = 0$ is equivalent to $\dfrac{\sin(4x+x)}{\cos 4x \cos x} = 0$.

It follows that $\sin 5x = 0$.

Hence, $5x = k\pi$ or $x = \dfrac{k\pi}{5}$, $k \in \mathbb{Z}$.

Problem 38. Solve the following equations:
1. $\sin^2 x + 2\sin x - 3 = 0$;
2. $2\cos^2 x + \cos x - 1 = 0$;
3. $\tan^2 x - \tan x - 2 = 0$.

Solution.   1. $\sin^2 x + 2\sin x - 3 = 0$
Let $t = \sin x$. Then $-1 \leq t \leq 1$.
The given equation is equivalent to $t^2 + 2t - 3 = 0$.
Since $a + b + c = 1 + 2 - 3 = 0$, then
$$t_1 = 1 \text{ and } t_2 = \frac{c}{a} = \frac{-3}{1} = -3.$$
We obtain $t = 1$ because $-1 \leq t \leq 1$.
It follows that $\sin x = 1$.
Hence, $x = \frac{\pi}{2} + k\pi$, where $k \in \mathbb{Z}$.

2. $2\cos^2 x + \cos x - 1 = 0$.
Let $t = \cos x$. Then $-1 \leq t \leq 1$.
The given equation can be rewritten as $2t^2 + t - 1 = 0$.
Then $(t + 1)(2t - 1) = 0$.
It follows that $\begin{bmatrix} t + 1 = 0 \\ 2t - 1 = 0 \end{bmatrix}$ or $\begin{bmatrix} t = -1 \\ t = \frac{1}{2} \end{bmatrix}$.

- If $t = -1$, we get $\cos x = -1$. Hence, $x = \pi + 2k\pi$, where $k \in \mathbb{Z}$.
- If $t = \frac{1}{2}$, we get $\cos x = \frac{1}{2}$. Hence, $x = \pm\frac{\pi}{3} + 2k\pi$, where $k \in \mathbb{Z}$.

3. $\tan^2 x - \tan x - 2 = 0$
The given equation can be written as
$$(\tan x + 1)(\tan x - 2) = 0.$$
Hence, $\begin{bmatrix} \tan x + 1 = 0 \\ \tan x - 2 = 0 \end{bmatrix}$ or $\begin{bmatrix} \tan x = -1 \\ \tan x = 2 \end{bmatrix}$.

It implies that $\begin{bmatrix} x = -\frac{\pi}{4} + k\pi \\ x = \theta + k\pi \end{bmatrix}$, where θ is an angle such that $\tan \theta = 2$.
Therefore, $x = \theta + k\pi$ or $x = -\frac{\pi}{4} + k\pi$, where $k \in \mathbb{Z}$.

Chapter 8. Solutions

Problem 39. Solve the following equations:

1. $2\cos^2 x - 3\sqrt{2}\cos x + 2 = 0$;
2. $\sin^2 x - 3\sin x + 2 = 0$;
3. $\cos^2 \dfrac{x}{2} - \cos \dfrac{x}{2} - 2 = 0$;
4. $\tan^3 x - 3\tan^2 x + 3\tan x - 1 = 0$;
5. $\dfrac{1}{\sin^2 x} = \cot x + 3$;
6. $4 - \cos 2x - 7\sin x = 0$;
7. $\tan^2 \dfrac{x}{2} - \left(1 - \sqrt{3}\right)\tan \dfrac{x}{2} - \sqrt{3} = 0$.

Solution. Solve the equations:

1. $2\cos^2 x - 3\sqrt{2}\cos x + 2 = 0$
   Let $t = \cos x$. Then $-1 \leq t \leq 1$.
   The given equation is equivalent to $2t^2 - 3\sqrt{2}t + 2 = 0$.
   The discriminant of the last equation is
   $$\Delta = b^2 - 4ac = \left(3\sqrt{2}\right)^2 - 4(2)(2) = 18 - 16 = 2.$$
   It follows that $t_1 = \dfrac{-b + \sqrt{\Delta}}{2a} = \dfrac{3\sqrt{2} + \sqrt{2}}{2(2)} = \sqrt{2}$
   and $t_2 = \dfrac{-b - \sqrt{\Delta}}{2a} = \dfrac{3\sqrt{2} - \sqrt{2}}{2(2)} = \dfrac{\sqrt{2}}{2}$.
   Since $-1 \leq t \leq 1$, we obtain $t = \dfrac{\sqrt{2}}{2}$.
   It implies that $\cos x = \dfrac{\sqrt{2}}{2} = \cos \dfrac{\pi}{4}$.
   Therefore, $x = \pm \dfrac{\pi}{4} + 2k\pi$, where $k \in \mathbb{Z}$.

2. $\sin^2 x - 3\sin x + 2 = 0$;
   We have $\sin^2 x - 3\sin x + 2 = 0$. Let $t = \sin x$.
   Then $-1 \leq t \leq 1$.
   The given equation can be written as
   $$t^2 - 3t + 2 = 0$$

or
$$(t-1)(t-2) = 0.$$
Then $t - 1 = 0$ or $t = 1$.
It turns out that $\sin x = 1$.
Therefore, $x = \dfrac{\pi}{2} + 2k\pi$, where $k \in \mathbb{Z}$.

3. $\cos^2 \dfrac{x}{2} - \cos \dfrac{x}{2} - 2 = 0$;
Let $t = \cos \dfrac{x}{2}$. Then $-1 \leq t \leq 1$. The given equation can be written as
$$t^2 - t - 2 = 0$$
or
$$(t+1)(t-2) = 0.$$
Hence, $t + 1 = 0$ or $t = -1$.
It turns out that $\sin x = -1$.
Therefore, $x = -\dfrac{\pi}{2} + 2k\pi$, where $k \in \mathbb{Z}$.

4. $\tan^3 x - 3\tan^2 x + 3\tan x - 1 = 0$;
We know that $(a-b)^3 = a^3 - 3a^2b + 3ab^2 - b^3$ for all numbers a and b. Then the given equation can be written as
$$(\tan x - 1)^3 = 0.$$
It follows that $\tan x - 1 = 0$. Then $\tan x = 1$.
Therefore, $x = \dfrac{\pi}{4} + k\pi$, where $k \in \mathbb{Z}$.

5. $\dfrac{1}{\sin^2 x} = \cot x + 3$
The equation is well defined when $\sin x \neq 0$ or $x \neq k\pi$.
Since $1 + \cot^2 x = \dfrac{1}{\sin^2 x}$, the given equation can be written as
$$1 + \cot^2 x = \cot x + 3 \quad \text{or} \quad \cot^2 x - \cot x - 2 = 0.$$
Solve the last equation, we obtain $\cot x = -1$ or $\cot x = 2$.

- If $\cot x = -1 = \cot\left(-\dfrac{\pi}{4}\right)$, then $x = -\dfrac{\pi}{4} + k\pi$, where $k \in \mathbb{Z}$.
- If $\cot x = 2 = \cot \alpha$, then $x = \alpha + k\pi$, where $k \in \mathbb{Z}$.

6. $4 - \cos 2x - 7 \sin x = 0$
Using double formula, $\cos 2x = 1 - 2 \sin^2 x$, we obtain

$$4 - \left(1 - 2\sin^2 x\right) - 7 \sin x = 0 \quad \text{or} \quad 2\sin^2 x - 7 \sin x + 3 = 0.$$

Let $t = \sin x$. Then $-1 \le t \le 1$.
The given equation can be written as $2t^2 - 7t + 3 = 0$.
The discriminant of the last equation is

$$\Delta = b^2 - 4ac = 7^2 - 4\,(2)\,(3) = 47 - 24 = 25.$$

It follows that

$$t_1 = \frac{-b + \sqrt{\Delta}}{2a} = \frac{7 + \sqrt{25}}{2\,(2)} = \frac{7 + 5}{4} = 3$$

and

$$t_2 = \frac{-b - \sqrt{\Delta}}{2a} = \frac{7 - \sqrt{25}}{2\,(2)} = \frac{7 - 5}{4} = \frac{1}{2}.$$

Since $-1 \le t \le 1$, we obtain $t = \frac{1}{2}$.
It implies that $\sin x = \frac{1}{2} = \sin \frac{\pi}{6}$.

Therefore, $\begin{bmatrix} x = \frac{\pi}{6} + 2k\pi \\ x = \pi - \frac{\pi}{6} + 2k\pi \end{bmatrix}$ or $\begin{bmatrix} x = \frac{\pi}{6} + 2k\pi \\ x = \frac{5\pi}{6} + 2k\pi \end{bmatrix}$, where $k \in \mathbb{Z}$.

7. $\tan^2 \frac{x}{2} - \left(1 - \sqrt{3}\right) \tan \frac{x}{2} - \sqrt{3} = 0$.
Let $t = \tan \frac{x}{2}$. The given equation is equivalent to

$$t^2 - \left(1 - \sqrt{3}\right) t - \sqrt{3} = 0.$$

Observe that $a + b + c = 1 - \left(1 - \sqrt{3}\right) - \sqrt{3} = 0$.
We obtain $t_1 = 1$ and $t_2 = \frac{c}{a} = -\sqrt{3}$.

- If $t = 1$, then $\tan \frac{x}{2} = 1 = \tan \frac{\pi}{4}$. It follows that

$$\frac{x}{2} = \frac{\pi}{4} + k\pi \quad \text{or} \quad x = \frac{\pi}{2} + 2k\pi.$$

- If $t = -\sqrt{3}$, then $\tan \dfrac{x}{2} = -\sqrt{3} = \tan\left(-\dfrac{\pi}{3}\right)$.
  It implies that $\dfrac{x}{2} = -\dfrac{\pi}{3} + k\pi$ or $x = -\dfrac{2\pi}{3} + 2k\pi$.

Thus, $x = -\dfrac{2\pi}{3} + 2k\pi$ or $x = \dfrac{\pi}{2} + 2k\pi$, where $k \in \mathbb{Z}$.

Problem 40. Solve the following equations:

1. $\sin x + \sqrt{3} \cos x = 1$;

2. $\sin x + \cos x = 1$;

3. $\cos 2x - \sin 2x = -1$;

4. $\sqrt{3} \sin x + \cos x = \sqrt{2}$;

5. $\cos x - \sqrt{3} \sin x = 3$.

Solution. Solve the equations:

1. $\sin x + \sqrt{3}\cos x = 1$
   We have $\sin x + \sqrt{3}\cos x = 1$.
   Multiply both sides of the equation by $\dfrac{1}{2}$, we obtain
   $$\dfrac{1}{2}\sin x + \dfrac{\sqrt{3}}{2}\cos x = \dfrac{1}{2}$$
   or
   $$\sin x \cos \dfrac{\pi}{3} + \sin \dfrac{\pi}{3} \cos x = \dfrac{1}{2}.$$
   Then $\sin\left(x + \dfrac{\pi}{3}\right) = \sin \dfrac{\pi}{6}$.
   It follows that
   $$\left[\begin{array}{l} x + \dfrac{\pi}{3} = \dfrac{\pi}{6} + 2k\pi \\ x + \dfrac{\pi}{3} = \pi - \dfrac{\pi}{6} + 2k\pi \end{array}\right. \text{ or } \left[\begin{array}{l} x = -\dfrac{\pi}{6} + 2k\pi \\ x = \dfrac{\pi}{2} + 2k\pi \end{array}\right..$$
   Hence, $x = -\dfrac{\pi}{6} + 2k\pi$ or $x = \dfrac{\pi}{2} + 2k\pi$, where $k \in \mathbb{Z}$.

2. $\sin x + \cos x = 1$
   We have $\sin x + \cos x = 1$. Multiply both sides of the equation by $\dfrac{\sqrt{2}}{2}$, we obtain
   $$\dfrac{\sqrt{2}}{2}\sin x + \dfrac{\sqrt{2}}{2}\cos x = \dfrac{\sqrt{2}}{2}$$

or
$$\cos x \cos \frac{\pi}{4} + \sin x \sin \frac{\pi}{4} = \cos \frac{\pi}{4}.$$

Then $\cos\left(x - \frac{\pi}{4}\right) = \cos\frac{\pi}{4}$.

It follows that $\left[\begin{array}{l} x - \frac{\pi}{4} = -\frac{\pi}{4} + 2k\pi \\ x - \frac{\pi}{4} = \frac{\pi}{4} + 2k\pi \end{array}\right.$. Then $\left[\begin{array}{l} x = 2k\pi \\ x = \frac{\pi}{2} + 2k\pi \end{array}\right.$.

Therefore, $x = 2k\pi$ or $x = \frac{\pi}{2} + 2k\pi$, where $k \in \mathbb{Z}$.

3. $\cos 2x - \sin 2x = -1$

We have $\cos 2x - \sin 2x = -1$.

Multiply both sides of the equation by $\frac{\sqrt{2}}{2}$, we get

$$\frac{\sqrt{2}}{2}\cos 2x - \frac{\sqrt{2}}{2}\sin 2x = -\frac{\sqrt{2}}{2}$$

or
$$\cos 2x \cos \frac{\pi}{4} - \sin 2x \sin \frac{\pi}{4} = \cos \frac{3\pi}{4}.$$

Then $\cos\left(2x + \frac{\pi}{4}\right) = \cos\frac{3\pi}{4}$.

We obtain $\left[\begin{array}{l} 2x + \frac{\pi}{4} = \frac{3\pi}{4} + 2k\pi \\ 2x + \frac{\pi}{4} = -\frac{3\pi}{4} + 2k\pi \end{array}\right.$ or $\left[\begin{array}{l} 2x = \frac{\pi}{2} + 2k\pi \\ 2x = -\pi + 2k\pi \end{array}\right.$.

Hence, $\left[\begin{array}{l} x = \frac{\pi}{4} + k\pi \\ x = -\frac{\pi}{2} + k\pi \end{array}\right.$.

Consequently, $x = \frac{\pi}{4} + k\pi$ or $x = -\frac{\pi}{2} + k\pi$, where $k \in \mathbb{Z}$.

4. $\sqrt{3}\sin x + \cos x = \sqrt{2}$

We have $\sqrt{3}\sin x + \cos x = \sqrt{2}$.

Multiply both sides of the equation by $\frac{1}{2}$, it implies that

$$\frac{\sqrt{3}}{2}\sin x + \frac{1}{2}\cos x = \frac{\sqrt{2}}{2}$$

or
$$\sin x \cos \frac{\pi}{6} + \sin \frac{\pi}{6} \cos x = \sin \frac{\pi}{4}.$$

Then $\sin\left(x+\dfrac{\pi}{6}\right) = \sin\dfrac{\pi}{4}$.

It follows that $\begin{bmatrix} x+\dfrac{\pi}{6} = \dfrac{\pi}{4} + 2k\pi \\ x+\dfrac{\pi}{6} = \pi - \dfrac{\pi}{4} + 2k\pi \end{bmatrix}$ or $\begin{bmatrix} x = \dfrac{\pi}{12} + 2k\pi \\ x = \dfrac{7\pi}{12} + 2k\pi \end{bmatrix}$.

Therefore, $x = \dfrac{\pi}{12} + 2k\pi$ or $x = \dfrac{7\pi}{12} + 2k\pi$, where $k \in \mathbb{Z}$.

5. $\cos x - \sqrt{3}\sin x = 3$.

We have $\cos x - \sqrt{3}\sin x = 3$.

Multiply both sides of the equation by $\dfrac{1}{2}$, we obtain

$$\dfrac{1}{2}\cos x - \dfrac{\sqrt{3}}{2}\sin x = \dfrac{3}{2}$$

or

$$\cos x \cos\dfrac{\pi}{3} - \sin x \sin\dfrac{\pi}{3} = \dfrac{3}{2}.$$

Then $\cos\left(x+\dfrac{\pi}{3}\right) = \dfrac{3}{2}$.

Since $\cos\left(x+\dfrac{\pi}{3}\right) \leq 1$, then there is no x that satisfies the equation.

Hence, the equation has no roots.

Problem 41. Solve the following equations:

1. $\sin^2 x + \sqrt{3}\sin x \cos x + 2\cos^2 x = 2$;

2. $3\sin^2 x - 2\sin x \cos x + 5\cos^2 x = 3$.

Solution. Solve the equations:

1. $\sin^2 x + \sqrt{3}\sin x \cos x + 2\cos^2 x = 2$

    - If $\cos x = 0$ or $x = \dfrac{\pi}{2} + k\pi$, the given equation is not satisfied.
    - If $\cos x \neq 0$ or $x \neq \dfrac{\pi}{2} + k\pi$, we divide both sides of the equation by $\cos^2 x$, implying that

    $$\dfrac{\sin^2 x}{\cos^2 x} + \dfrac{\sqrt{3}\sin x \cos x}{\cos^2 x} + \dfrac{2\cos^2 x}{\cos^2 x} = \dfrac{2}{\cos^2 x}$$

    or

    $$\tan^2 x + \sqrt{3}\tan x + 2 = 2\left(1+\tan^2 x\right).$$

Chapter 8. Solutions

Then $\tan^2 x - \sqrt{3}\tan x = 0$ or $\tan x\left(\tan x - \sqrt{3}\right) = 0$.

It follows that $\begin{bmatrix} \tan x = 0 \\ \tan x - \sqrt{3} = 0 \end{bmatrix}$ or $\begin{bmatrix} \tan x = 0 \\ \tan x = \sqrt{3} = \tan\dfrac{\pi}{3} \end{bmatrix}$.

Hence, $x = k\pi$ or $x = \dfrac{\pi}{3} + k\pi$, where $k \in \mathbb{Z}$.

2. $3\sin^2 x - 2\sin x \cos x + 5\cos^2 x = 3$.
If $\cos x = 0$ or $x = \dfrac{\pi}{2} + k\pi$, the given equation is not satisfied.
If $\cos x \neq 0$ or $x \neq \dfrac{\pi}{2} + k\pi$, we divide both sides of the equation by $\cos^2 x$, implying that

$$\frac{3\sin^2 x}{\cos^2 x} - \frac{2\sin x \cos x}{\cos^2 x} + \frac{5\cos^2 x}{\cos^2 x} = \frac{3}{\cos^2 x}$$

or

$$3\tan^2 x - 2\tan x + 5 = 3\left(1 + \tan^2 x\right).$$

Then $\tan x = 1 = \tan\dfrac{\pi}{4}$.

Consequently, $x = \dfrac{\pi}{4} + k\pi$, where $k \in \mathbb{Z}$.

Problem 42. Let a, b, c be real numbers different from -1 and 1 and satisfy
$$a + b + c = abc.$$

Show that $\dfrac{a}{1-a^2} + \dfrac{b}{1-b^2} + \dfrac{c}{1-c^2} = \dfrac{4abc}{(1-a^2)(1-b^2)(1-c^2)}.$

Solution. Since a, b, c are real numbers, all different from -1 and 1, such that $a + b + c = abc$, then there exist a triangle ABC that satisfies $a = \tan A$, $b = \tan B$ and $C = \tan C$.
We obtain

$$\frac{a}{1-a^2} + \frac{b}{1-b^2} + \frac{c}{1-c^2} = \frac{\tan A}{1-\tan^2 A} + \frac{\tan B}{1-\tan^2 B} + \frac{\tan C}{1-\tan^2 C}$$

$$= \frac{1}{2}\left(\frac{2\tan A}{1-\tan^2 A} + \frac{2\tan B}{1-\tan^2 B} + \frac{2\tan C}{1-\tan^2 C}\right)$$

$$= \frac{1}{2}\left(\tan 2A + \tan 2B + \tan 2C\right).$$

Moreover, $A + B + C = \pi$ or $2A + 2B = 2\pi - 2C$. Then
$$\tan(2A + 2B) = \tan(2\pi - 2C) \text{ or } \frac{\tan 2A + \tan 2B}{1 - \tan 2A \tan 2B} = -\tan 2C.$$
It implies that $\tan 2A + \tan 2B = -\tan 2C + \tan 2A \tan 2B \tan 2C$.
As a result, $\tan 2A + \tan 2B + \tan 2C = \tan 2A \tan 2B \tan 2C$.
We obtain
$$\frac{a^2}{1-a^2} + \frac{b^2}{1-b^2} + \frac{c^2}{1-c^2}$$
$$= \frac{1}{2}(\tan 2A \tan 2B \tan 2C)$$
$$= \frac{1}{2}\left(\frac{2\tan A}{1-\tan^2 A}\right)\left(\frac{2\tan B}{1-\tan^2 B}\right)\left(\frac{2\tan C}{1-\tan^2 C}\right)$$
$$= \frac{4\tan A \tan B \tan C}{(1-\tan^2 A)(1-\tan^2 B)(1-\tan^2 C)}$$
$$= \frac{4abc}{(1-a^2)(1-b^2)(1-c^2)}.$$
Thus, the claim is proved.

Problem 43. Given real numbers x, y and z such that x, y and $z > 0$. Prove that
$$\frac{x}{x + \sqrt{(x+y)(x+z)}} + \frac{y}{y + \sqrt{(y+z)(y+x)}} + \frac{z}{z + \sqrt{(z+x)(z+y)}} \leq 1.$$

Solution. Prove that $\dfrac{x}{x + \sqrt{(x+y)(x+z)}} + \dfrac{y}{y + \sqrt{(y+z)(y+x)}}$
$+ \dfrac{z}{z + \sqrt{(z+x)(z+y)}} \leq 1.$

Since the given inequality is homogeneous, Without loss of generality, suppose that $xy + yz + zx = 1$. Then there exists a triangle ABC such that $x = \tan\dfrac{A}{2}, y = \tan\dfrac{B}{2}$ and $z = \tan\dfrac{C}{2}$.

We have $\dfrac{x}{x + \sqrt{(x+y)(x+z)}} = \dfrac{1}{1 + \sqrt{\dfrac{(x+y)(x+z)}{x^2}}}.$

Moreover,
$$\frac{(x+y)(x+z)}{x^2} = \frac{\left(\tan\dfrac{A}{2} + \tan\dfrac{B}{2}\right)\left(\tan\dfrac{A}{2} + \tan\dfrac{C}{2}\right)}{\tan^2\dfrac{A}{2}}$$

$$= \frac{\sin\left(\frac{A+B}{2}\right)\sin\left(\frac{A+C}{2}\right)}{\cos^2\frac{A}{2}\cos\frac{B}{2}\cos\frac{C}{2}\tan^2\frac{A}{2}}$$

$$= \frac{\cos\frac{C}{2}\cos\frac{B}{2}}{\cos\frac{B}{2}\cos\frac{C}{2}\sin^2\frac{A}{2}}$$

$$= \frac{1}{\sin^2\frac{A}{2}}.$$

Then $\dfrac{x}{x+\sqrt{(x+y)(x+z)}} = \dfrac{1}{1+\sqrt{\dfrac{1}{\sin^2\frac{A}{2}}}} = \dfrac{1}{1+\dfrac{1}{\sin\frac{A}{2}}} = \dfrac{\sin\frac{A}{2}}{1+\sin\frac{A}{2}}.$

Similarly,

$$\frac{y}{y+\sqrt{(y+z)(y+x)}} = \frac{\sin\frac{B}{2}}{1+\sin\frac{B}{2}}$$

and

$$\frac{z}{z+\sqrt{(z+x)(z+y)}} = \frac{\sin\frac{C}{2}}{1+\sin\frac{C}{2}}.$$

Using Cauchy-Schwarz inequality and $\sin\frac{A}{2}+\sin\frac{B}{2}+\sin\frac{C}{2} \geq \frac{3}{2}$, we obtain

$$\frac{x}{x+\sqrt{(x+y)(x+z)}} + \frac{y}{y+\sqrt{(y+z)(y+x)}} + \frac{z}{z+\sqrt{(z+x)(z+y)}}$$

$$= \frac{\sin\frac{A}{2}}{1+\sin\frac{A}{2}} + \frac{\sin\frac{B}{2}}{1+\sin\frac{B}{2}} + \frac{\sin\frac{C}{2}}{1+\sin\frac{C}{2}}$$

$$= 3 - \left(\frac{1}{1+\sin\frac{A}{2}} + \frac{1}{1+\sin\frac{B}{2}} + \frac{1}{1+\sin\frac{C}{2}}\right)$$

$$\leq 3 - \frac{(1+1+1)^2}{3 + \sin\frac{A}{2} + \sin\frac{B}{2} + \sin\frac{C}{2}}$$

$$\leq 3 - \frac{9}{3 + \frac{3}{2}} = 3 - 2 = 1.$$

Therefore, $\dfrac{x}{x + \sqrt{(x+y)(x+z)}} + \dfrac{y}{y + \sqrt{(y+z)(y+x)}}$

$+ \dfrac{z}{z + \sqrt{(z+x)(z+y)}} \leq 1.$

Problem 44. Suppose that x, y and z are positive real numbers that satisfy $x + y + z = xyz$.
Prove that

1. $\dfrac{1}{\sqrt{1+x^2}} + \dfrac{1}{\sqrt{1+y^2}} + \dfrac{1}{\sqrt{1+z^2}} \leq \dfrac{3}{2};$

2. $\dfrac{x}{\sqrt{1+x^2}} + \dfrac{y}{\sqrt{1+y^2}} + \dfrac{z}{\sqrt{1+z^2}} \leq \dfrac{3\sqrt{3}}{2}.$

Solution. Since $x, y, z > 0$ and satisfy $x+y+z = xyz$, then there exists an acute triangle ABC such that $x = \tan A, y = \tan B$ and $z = \tan C$.
Prove that

1. $\dfrac{1}{\sqrt{1+x^2}} + \dfrac{1}{\sqrt{1+y^2}} + \dfrac{1}{\sqrt{1+z^2}} \leq \dfrac{3}{2}$

We have $\dfrac{1}{\sqrt{1+x^2}} + \dfrac{1}{\sqrt{1+y^2}} + \dfrac{1}{\sqrt{1+z^2}}$

$= \dfrac{1}{\sqrt{1+\tan^2 A}} + \dfrac{1}{\sqrt{1+\tan^2 B}} + \dfrac{1}{\sqrt{1+\tan^2 C}}$

$= \dfrac{1}{\sqrt{\dfrac{1}{\cos^2 A}}} + \dfrac{1}{\sqrt{\dfrac{1}{\cos^2 B}}} + \dfrac{1}{\sqrt{\dfrac{1}{\cos^2 C}}}$

$= \dfrac{1}{\dfrac{1}{\cos A}} + \dfrac{1}{\dfrac{1}{\cos B}} + \dfrac{1}{\dfrac{1}{\cos C}}$

$= \cos A + \cos B + \cos C.$

Moreover, $\cos A + \cos B + \cos C \leq \dfrac{3}{2}$.

Consequently, $\dfrac{1}{\sqrt{1+x^2}} + \dfrac{1}{\sqrt{1+y^2}} + \dfrac{1}{\sqrt{1+z^2}} \leq \dfrac{3}{2}$.

2. $\dfrac{x}{\sqrt{1+x^2}} + \dfrac{y}{\sqrt{1+y^2}} + \dfrac{z}{\sqrt{1+z^2}} \leq \dfrac{3\sqrt{3}}{2}$

We have
$$\dfrac{x}{\sqrt{1+x^2}} + \dfrac{y}{\sqrt{1+y^2}} + \dfrac{z}{\sqrt{1+z^2}}$$
$$= \dfrac{\tan A}{\sqrt{1+\tan^2 A}} + \dfrac{\tan B}{\sqrt{1+\tan^2 B}} + \dfrac{\tan C}{\sqrt{1+\tan^2 C}}$$
$$= \dfrac{\tan A}{\sqrt{\dfrac{1}{\cos^2 A}}} + \dfrac{\tan B}{\sqrt{\dfrac{1}{\cos^2 B}}} + \dfrac{\tan C}{\sqrt{\dfrac{1}{\cos^2 C}}}$$
$$= \tan A \cos A + \tan B \cos B + \tan C \cos C$$
$$= \sin A + \sin B + \sin C.$$

Since $\sin A + \sin B + \sin C \leq \dfrac{3\sqrt{3}}{2}$, we obtain

$$\dfrac{x}{\sqrt{1+x^2}} + \dfrac{y}{\sqrt{1+y^2}} + \dfrac{z}{\sqrt{1+z^2}} \leq \dfrac{3\sqrt{3}}{2}.$$

Therefore, $\dfrac{x}{\sqrt{1+x^2}} + \dfrac{y}{\sqrt{1+y^2}} + \dfrac{z}{\sqrt{1+z^2}} \leq \dfrac{3\sqrt{3}}{2}$.

Problem 45. Solve the following system of equations:
$$\begin{cases} x + y = \dfrac{\pi}{2} \\ \sin x + \sin y = 1 \end{cases}.$$

Solution. Solve the equation system:
We have
$$x + y = \dfrac{\pi}{2} \qquad (1)$$
$$\sin x + \sin y = 1 \qquad (2)$$

Using sum to product formula, from (2), we obtain
$$2 \sin\left(\dfrac{x+y}{2}\right) \cos\left(\dfrac{x-y}{2}\right) = 1.$$

Moreover, $x+y = \dfrac{\pi}{2}$, it follows that

$$2\sin\dfrac{\pi}{4}\cos\left(\dfrac{x-y}{2}\right) = 1 \quad \text{or} \quad \cos\left(\dfrac{x-y}{2}\right) = \dfrac{\sqrt{2}}{2}.$$

We obtain
$$x - y = \pm\dfrac{\pi}{4} + 2k\pi$$
, where $k \in \mathbb{Z}$.

- If $x - y = \dfrac{\pi}{4} + 2k\pi$, we obtain $\begin{cases} x - y = \dfrac{\pi}{4} + 2k\pi \\ x + y = \dfrac{\pi}{2} \end{cases}$.

  Adding both equations, we obtain
  $$2x = \dfrac{\pi}{4} + \dfrac{\pi}{2} + 2k\pi$$
  $$= \dfrac{\pi}{4} + \dfrac{2\pi}{4} + 2k\pi$$
  $$= \dfrac{3\pi}{4} + 2k\pi.$$

  It implies that $x = \dfrac{3\pi}{8} + k\pi$. Then
  $$y = \dfrac{\pi}{2} - x$$
  $$= \dfrac{\pi}{2} - \left(\dfrac{3\pi}{8} + k\pi\right)$$
  $$= \dfrac{\pi}{2} - \dfrac{3\pi}{8} - k\pi$$
  $$= \dfrac{4\pi}{8} - \dfrac{3\pi}{8} - k\pi$$
  $$= \dfrac{\pi}{8} - k\pi.$$

  Therefore, $x = \dfrac{3\pi}{8} + k\pi$ and $y = \dfrac{\pi}{8} - k\pi$, where $k \in \mathbb{Z}$.

- If $x - y = -\dfrac{\pi}{4} + 2k\pi$, we obtain $\begin{cases} x - y = -\dfrac{\pi}{4} + 2k\pi \\ x + y = \dfrac{\pi}{2} \end{cases}$.

  Adding both equations, we obtain
  $$2x = -\dfrac{\pi}{4} + \dfrac{\pi}{2} + 2k\pi$$

$$= -\frac{\pi}{4} + \frac{2\pi}{4} + 2k\pi$$
$$= \frac{\pi}{4} + 2k\pi.$$

It implies that $x = \frac{\pi}{8} + k\pi$. Then

$$y = \frac{\pi}{2} - x$$
$$= \frac{\pi}{2} - \left(\frac{\pi}{8} + k\pi\right)$$
$$= \frac{\pi}{2} - \frac{\pi}{8} - k\pi$$
$$= \frac{4\pi}{8} - \frac{\pi}{8} - k\pi$$
$$= \frac{3\pi}{8} - k\pi.$$

Therefore, $x = \frac{\pi}{8} + k\pi$ and $y = \frac{3\pi}{8} - k\pi$, where $k \in \mathbb{Z}$.

Problem 46. Solve the following system of equations:

$$\begin{cases} x + y = \dfrac{\pi}{2} \\ \tan x + \tan y = 2 \end{cases}.$$

Solution. Solve the equation system:
We have

$$x + y = \frac{\pi}{2} \qquad (1)$$
$$\tan x + \tan y = 2 \qquad (2)$$

The system of equatons is well-defined if and only $x, y \neq \frac{\pi}{2} + k\pi$, where $k \in \mathbb{Z}$.
From (1): $x + y = \frac{\pi}{2}$, then $y = \frac{\pi}{2} - x$.
Plug $y = \frac{\pi}{2} - x$ in (2), it follows that

$$\tan x + \tan\left(\frac{\pi}{2} - x\right) = 2$$

or

$$\tan x + \frac{1}{\tan x} - 2 = 0.$$

170

Multiply both sides of the equation by $\tan x$, we obtain
$$\tan^2 x + 1 - 2\tan x = 0$$
or
$$(\tan x - 1)^2 = 0.$$
It follows that $\tan x - 1 = 0$. Then $\tan x = 1$.
Hence, $x = \dfrac{\pi}{4} + k\pi$, where $k \in \mathbb{Z}$.
It implies that
$$\begin{aligned} y &= \frac{\pi}{2} - x \\ &= \frac{\pi}{2} - \left(\frac{\pi}{4} + k\pi\right) \\ &= \frac{\pi}{2} - \frac{\pi}{4} - k\pi \\ &= \frac{2\pi}{4} - \frac{\pi}{4} - k\pi \\ &= \frac{\pi}{4} - k\pi. \end{aligned}$$

Therefore, $x = \dfrac{\pi}{4} + k\pi$ and $y = \dfrac{\pi}{4} - k\pi$, where $k \in \mathbb{Z}$.

Problem 47. Let x, y and z be real numbers such that $\sin x + \sin y + \sin z = 0$ and $\cos x + \cos y + \cos z = 0$. Prove that

1. $\cos(x - y) = -\dfrac{1}{2}$;

2. $\cos(\theta - x) + \cos(\theta - y) + \cos(\theta - z) = 0$ for all $\theta \in \mathbb{R}$;

3. $\sin^2 x + \sin^2 y + \sin^2 z = \dfrac{3}{2}$;

4. $2\left(\cot^2 x \cot^2 y + \cot^2 y \cot^2 z + \cot^2 z \cot^2 x\right) = 9\cot^2 x \cot^2 y \cot^2 z$.

Solution. Prove that

1. $\cos(x - y) = -\dfrac{1}{2}$

We have $\sin x + \sin y + \sin z = 0$ or $\sin x + \sin y = -\sin z$.
Then $(\sin x + \sin y)^2 = (-\sin z)^2$ or
$$\sin^2 x + 2\sin x \sin y + \sin^2 y = \sin^2 z. \qquad (1)$$

171

Moreover, $\cos x + \cos y + \cos z = 0$ or $\cos x + \cos y = -\cos z$.
It follows that $(\cos x + \cos y)^2 = (-\cos z)^2$ or

$$\cos^2 x + 2\cos x \cos y + \cos^2 y = \cos^2 z. \qquad (2)$$

Adding (1) and (2), we obtain

$$1 + 2(\cos x \cos y + \sin x \sin y) + 1 = 1$$

or

$$2\cos(x - y) = -1.$$

Consequently, $\cos(x - y) = -\dfrac{1}{2}$.

2. $\cos(\theta - x) + \cos(\theta - y) + \cos(\theta - z) = 0$ for all $\theta \in \mathbb{R}$
Using difference formula, we have

$$\cos(\theta - x) = \cos\theta \cos x + \sin\theta \sin x$$
$$\cos(\theta - y) = \cos\theta \cos y + \sin\theta \sin y$$
and $\quad \cos(\theta - z) = \cos\theta \cos z + \sin\theta \sin z.$

Using the fact that $\cos x + \cos y + \cos z = 0$ and $\sin x + \sin y + \sin z = 0$, it implies that

$$\cos(\theta - x) + \cos(\theta - y) + \cos(\theta - z)$$
$$= \cos\theta(\cos x + \cos y + \cos z) + \sin\theta(\sin x + \sin y + \sin z)$$
$$= \cos\theta(0) + \sin\theta(0) = 0$$

Therefore, $\cos(\theta - x) + \cos(\theta - y) + \cos(\theta - z) = 0$.

3. $\sin^2 x + \sin^2 y + \sin^2 z = \dfrac{3}{2}$.
From (2), $\cos(\theta - x) + \cos(\theta - y) + \cos(\theta - z) = 0$ for all $\theta \in \mathbb{R}$.
Taking $\theta = x + y + z$, we obtain

$$\cos(y + z) + \cos(z + x) + \cos(x + y) = 0$$

or

$$\cos y \cos z - \sin y \sin z + \cos z \cos x - \sin z \sin x + \cos x \cos y$$
$$- \sin x \sin y = 0.$$

Then
$$\cos x \cos y + \cos y \cos z + \cos z \cos x$$
$$= \sin x \sin y + \sin y \sin z + \sin z \sin x. \quad (1)$$

Moreover, $\cos x + \cos y + \cos z = 0$.
Squaring both sides of the equality, it implies that

$$\cos^2 x + \cos^2 y + \cos^2 z + 2\left(\cos x \cos y + \cos y \cos z + \cos z \cos x\right) = 0.$$

From (1), we obtain

$$\cos^2 x + \cos^2 y + \cos^2 z + 2\left(\sin x \sin y + \sin y \sin z + \sin z \sin x\right) = 0. \quad (2)$$

Similarly,

$$\sin^2 x + \sin^2 y + \sin^2 z + 2\left(\sin x \sin y + \sin y \sin z + \sin z \sin x\right) = 0. \quad (3)$$

From (2) and (3), it follows that

$$\sin^2 x + \sin^2 y + \sin^2 z = \cos^2 x + \cos^2 y + \cos^2 z.$$

Using the fact that $\cos^2 \alpha = 1 - \sin^2 \alpha$, we obtain

$$\sin^2 x + \sin^2 y + \sin^2 z = 1 - \sin^2 x + 1 - \sin^2 y + 1 - \sin^2 z$$

or
$$2\left(\sin^2 x + \sin^2 y + \sin^2 z\right) = 3.$$

Consequently, $\sin^2 x + \sin^2 y + \sin^2 z = \dfrac{3}{2}$.

4. $2\left(\cot^2 x \cot^2 y + \cot^2 y \cot^2 z + \cot^2 z \cot^2 x\right) = 9\cot^2 x \cot^2 y \cot^2 z$.
Using the fact that $1 + \sin^2 \alpha = \dfrac{1}{\cot^2 \alpha}$, we obtain

$$\left(1 + \sin^2 x\right) + \left(1 + \sin^2 y\right) + \left(1 + \sin^2 z\right) = \dfrac{1}{\cot^2 x} + \dfrac{1}{\cot^2 y} + \dfrac{1}{\cot^2 z}.$$

Hence,
$$3 + \sin^2 x + \sin^2 y + \sin^2 z = \dfrac{1}{\cot^2 x} + \dfrac{1}{\cot^2 y} + \dfrac{1}{\cot^2 z}.$$

Chapter 8. Solutions

By knowing that $\sin^2 x + \sin^2 y + \sin^2 z = \dfrac{3}{2}$, it follows that

$$\frac{1}{\cot^2 x} + \frac{1}{\cot^2 y} + \frac{1}{\cot^2 z} = 3 + \frac{3}{2}$$
$$= \frac{6}{2} + \frac{3}{2}$$
$$= \frac{9}{2}.$$

Multiply both sides of the equality by $\cot^2 x \cot^2 y \cot^2 z$, we obtain $\cot^2 x \cot^2 y + \cot^2 y \cot^2 z + \cot^2 z \cot^2 x = \dfrac{9}{2}\cot^2 x \cot^2 y \cot^2 z$. Therefore, $2\left(\cot^2 x \cot^2 y + \cot^2 y \cot^2 z + \cot^2 z \cot^2 x\right)$.
$= 9\cot^2 x \cot^2 y \cot^2 z$.

Problem 48. Let (u_n) be a sequence such that $u_1 = \dfrac{\sqrt{2}}{2}$ and $u_{n+1} = \dfrac{\sqrt{2}}{2}\sqrt{1 - \sqrt{1 - u_n^2}}$. Find u_n in terms of n.

Solution. Find u_n in terms of n.
We have $u_1 = \dfrac{\sqrt{2}}{2} = \sin\dfrac{\pi}{2^2}$ and $u_{n+1} = \dfrac{\sqrt{2}}{2}\sqrt{1 - \sqrt{1 - u_n^2}}$.
Then

$$u_2 = \frac{\sqrt{2}}{2}\sqrt{1 - \sqrt{1 - u_1^2}}$$
$$= \frac{\sqrt{2}}{2}\sqrt{1 - \sqrt{1 - \sin^2 \frac{\pi}{2^2}}}$$
$$= \frac{\sqrt{2}}{2}\sqrt{1 - \sqrt{\cos^2 \frac{\pi}{2^2}}}$$
$$= \frac{\sqrt{2}}{2}\sqrt{1 - \cos\frac{\pi}{2^2}} = \frac{\sqrt{2}}{2}\sqrt{2\sin^2 \frac{\pi}{2^3}} = \sin^2 \frac{\pi}{2^3}.$$

Suppose that $u_n = \sin\dfrac{\pi}{2^{n+1}}$, we shall show that $u_{n+1} = \sin\dfrac{\pi}{2^{n+2}}$. Observe that

$$u_{n+1} = \frac{\sqrt{2}}{2}\sqrt{1 - \sqrt{1 - u_n^2}}$$

$$= \frac{\sqrt{2}}{2}\sqrt{1-\sqrt{1-\sin^2\frac{\pi}{2^{n+1}}}}$$

$$= \frac{\sqrt{2}}{2}\sqrt{1-\sqrt{\cos^2\frac{\pi}{2^{n+1}}}}$$

$$= \frac{\sqrt{2}}{2}\sqrt{1-\cos\frac{\pi}{2^{n+1}}}$$

$$= \frac{\sqrt{2}}{2}\sqrt{2\sin^2\frac{\pi}{2^{n+2}}} = \sin\frac{\pi}{2^{n+2}}.$$

Consequently, $u_n = \sin\frac{\pi}{2^{n+1}}$ for all $n \in \mathbb{N}$.

Problem 49. Suppose that $\tan x_1 \tan x_2 \ldots \tan x_n = k$. Find the maximum value of

$$A = \sin x_1 \sin x_2 \ldots \sin x_n.$$

Solution. Find the maximum value of A.
It is obviously to see that $\sin^2 x_1 + \cos^2 x_1 \geq 2\sin x_1 \cos x_1$.
Then $\sin x_1 \cos x_1 \leq \frac{1}{2}$.
Similarly,

$$\sin x_2 \cos x_2 \leq \frac{1}{2}$$
$$\sin x_3 \cos x_3 \leq \frac{1}{2}$$
$$\vdots$$
$$\text{and} \quad \sin x_n \cos x_n \leq \frac{1}{2}.$$

It follows that

$$\sin x_1 \sin x_2 \ldots \sin x_n \cos x_1 \cos x_2 \ldots \cos x_n \leq \frac{1}{2^n}. \qquad (1)$$

Since $\tan x_1 \tan x_2 \ldots \tan x_n = k$, it implies that

$$\frac{\sin x_1}{\cos x_1} \times \frac{\sin x_2}{\cos x_2} \times \ldots \times \frac{\sin x_n}{\cos x_n} = k.$$

Then $\cos x_1 \cos x_2 \ldots \cos x_n = \dfrac{1}{k} \sin x_1 \sin x_2 \ldots \sin x_n = \dfrac{1}{k} A$.
From (1), we obtain
$$A\left(\dfrac{1}{k}A\right) \leq \dfrac{1}{2^n}$$
or
$$A^2 \leq \dfrac{k}{2^n}.$$
It turns out that $A \leq \sqrt{\dfrac{k}{2^n}}$.

Thus, $\max A = \sqrt{\dfrac{k}{2^n}}$.

Problem 50. Compute $S = \tan^6 \dfrac{\pi}{18} + \tan^6 \dfrac{5\pi}{18} + \tan^6 \dfrac{7\pi}{18}$.

Solution. Compute S.
Observe that
$$\tan^2 3\left(\dfrac{\pi}{18}\right) = \tan^2 \dfrac{\pi}{6} = \left(\dfrac{\sqrt{3}}{3}\right)^2 = \dfrac{1}{3}$$
$$\tan^2 3\left(\dfrac{5\pi}{18}\right) = \tan^2 \left(\dfrac{5\pi}{6}\right) = \left(-\dfrac{\sqrt{3}}{3}\right)^2 = \dfrac{1}{3}$$
and $\tan^2 3\left(\dfrac{7\pi}{18}\right) = \tan^2 \dfrac{7\pi}{6} = \left(\dfrac{\sqrt{3}}{3}\right)^2 = \dfrac{1}{3}.$

Then $\dfrac{\pi}{18}, \dfrac{5\pi}{18}$ and $\dfrac{7\pi}{18}$ are roots of the equation $\tan^2 3x = \dfrac{1}{3}$.
Since $\tan 3x = \dfrac{3\tan x - \tan^3 x}{1 - 3\tan^2 x}$, it follows that
$$\left(\dfrac{3\tan x - \tan^3 x}{1 - 3\tan^2 x}\right)^2 = \dfrac{1}{3}$$
or
$$3\left(\tan^6 x - 6\tan^4 x + 9\tan^2 x\right) = 1 - 6\tan^2 x + 9\tan^4 x.$$

Then $3\tan^6 x - 27\tan^4 x + 33\tan^2 x - 1 = 0$.
Let $t = \tan^2 x$. We obtain
$$3t^3 - 27t^2 + 33t - 1 = 0. \tag{1}$$

Consequently, $t_1 = \tan^2 \frac{\pi}{18}, t_2 = \tan^2 \frac{5\pi}{18}$ and $t_3 = \tan^2 \frac{7\pi}{18}$ are the roots of the equation (1). Applying Vieta's theorem, we obtain

$$\begin{cases} t_1 + t_2 + t_3 = 9 \\ t_1 t_2 + t_2 t_3 + t_3 t_1 = 11 \\ t_1 t_2 t_3 = \frac{1}{3} \end{cases}.$$

Hence,

$$\begin{aligned} S &= t_1^3 + t_2^3 + t_3^3 \\ &= t_1^3 + t_2^3 + t_3^3 - 3t_1 t_2 t_3 + 3t_1 t_2 t_3 \\ &= (t_1 + t_2 + t_3)\left(t_1^2 + t_2^2 + t_3^2 - t_1 t_2 - t_2 t_3 - t_3 t_1\right) + 3t_1 t_2 t_3 \\ &= (t_1 + t_2 + t_3)\left[(t_1 + t_2 + t_3)^2 - 3(t_1 t_2 + t_2 t_3 + t_3 t_1)\right] + 3t_1 t_2 t_3 \\ &= 9\left[9^2 - 3(11)\right] + 3\left(\frac{1}{3}\right) = 433. \end{aligned}$$

Problem 51. Let (a_n) be a sequence defined by

$$a_n = \tan n° \tan (n-1)°.$$

Compute $S_n = \sum_{k=1}^{n} a_k$ in terms of n.

Solution. Compute $S_n = \sum_{k=1}^{n} a_k$ in terms of n.

Using difference formula, $\tan(a-b) = \dfrac{\tan a - \tan b}{1 + \tan a \tan b}$. We obtain

$$\tan\left[k° - (k-1)°\right] = \frac{\tan k° - \tan(k-1)°}{1 + \tan k° \tan(k-1)°}$$

or

$$\frac{\tan k° - \tan(k-1)°}{1 + \tan k° \tan(k-1)°} = \tan 1°.$$

It follows that

$$\tan k° \tan(k-1)° = \frac{1}{\tan 1°}\left[\tan k° - \tan(k-1)°\right] - 1.$$

Then $a_k = \dfrac{1}{\tan 1°} \left[\tan k° - \tan(k-1)°\right] - 1.$
Consequently,
$$S_n = \sum_{k=1}^{n} a_k = \sum_{k=1}^{n} \dfrac{1}{\tan 1°}\left[\tan k° - \tan(k-1)°\right] - 1$$
$$= \dfrac{1}{\tan 1°} \sum_{k=1}^{n}\left[\tan k° - \tan(k-1)°\right] - \sum_{k=1}^{n} 1$$
$$= \dfrac{1}{\tan 1°}(\tan n° - \tan 0°) - n = \dfrac{\tan n°}{\tan 1°} - n.$$

Therefore, $S_n = \dfrac{\tan n°}{\tan 1°} - n.$

Problem 52. Let
$$S_n = \tan\theta \tan^2\dfrac{\theta}{2} + 2\tan\dfrac{\theta}{2}\tan^2\dfrac{\theta}{2^2} + \ldots + 2^{n-1}\tan\dfrac{\theta}{2^{n-1}}\tan^2\dfrac{\theta}{2^n}.$$
Prove that $S_n = \tan\theta - 2^n \tan\dfrac{\theta}{2^n}.$

Solution. Prove that $S_n = \tan\theta - 2^n \tan\dfrac{\theta}{2^n}.$

Using double angle formula, $\tan 2\alpha = \dfrac{2\tan\alpha}{1-\tan^2\alpha}.$
Then
$$\tan 2\alpha - \tan 2\alpha \tan^2\alpha = 2\tan\alpha$$
or
$$\tan 2\alpha \tan^2\alpha = \tan 2\alpha - 2\tan\alpha.$$
It follows that
$$\tan\theta \tan^2\dfrac{\theta}{2} = \tan\theta - 2\tan\dfrac{\theta}{2}$$
$$2\tan\dfrac{\theta}{2}\tan^2\dfrac{\theta}{2^2} = 2\tan\dfrac{\theta}{2} - 2^2\tan\dfrac{\theta}{2^2}$$
$$\vdots$$
and $\quad 2^{n-1}\tan\dfrac{\theta}{2^{n-1}}\tan^2\dfrac{\theta}{2^n} = 2^{n-1}\tan\dfrac{\theta}{2^{n-1}} - 2^n\tan\dfrac{\theta}{2^n}$

Adding the equalities, we obtain $S_n = \tan\theta - 2^n \tan\dfrac{\theta}{2^n}.$

Therefore, $S_n = \tan\theta - 2^n\tan\dfrac{\theta}{2^n}.$

Problem 53. Let $a_1, a_2, ..., a_n$ be real numbers such that $0 < a_1 < a_2 < ... < a_n < \dfrac{\pi}{2}$. Prove that

$$\tan a_1 < \frac{\sin a_1 + \sin a_2 + ... + \sin a_n}{\cos a_1 + \cos a_2 + ... + \cos a_n} < \tan a_n.$$

Solution. Prove that $\tan a_1 < \dfrac{\sin a_1 + \sin a_2 + ... + \sin a_n}{\cos a_1 + \cos a_2 + ... + \cos a_n} < \tan a_n.$

From the hypothesis, $0 < a_1 < a_2 < ... < a_n < \dfrac{\pi}{2}$, since $\sin x$ is increasing on $\left(0, \dfrac{\pi}{2}\right)$ and $\cos x$ is decreasing on $\left(0, \dfrac{\pi}{2}\right)$, it follows that

$$0 < \sin a_1 < \sin a_2 < ... < \sin a_n$$

and

$$\cos a_1 > \cos a_2 > ... > \cos a_n > 0.$$

We obtain

$$\underbrace{\sin a_1 + \sin a_1 + ... + \sin a_1}_{n \text{ terms}} < \sin a_1 + \sin a_2 + ... + \sin a_n$$
$$< \underbrace{\sin a_n + \sin a_n + ... + \sin a_n}_{n \text{ terms}}$$

or

$$n \sin a_1 < \sin a_1 + \sin a_2 + ... + \sin a_n < n \sin a_n. \qquad (1)$$

Similarly, $n \cos a_n < \cos a_1 + \cos a_2 + ... + \cos a_n < n \cos a_1$. Then

$$\frac{1}{n \cos a_1} < \frac{1}{\cos a_1 + \cos a_2 + ... + \cos a_n} < \frac{1}{n \cos a_n}. \qquad (2)$$

Multiple (1) by (2), it implies that

$$\frac{n \sin a_1}{n \cos a_1} < \frac{\sin a_1 + \sin a_2 + ... + \sin a_n}{\cos a_1 + \cos a_2 + ... + \cos a_n} < \frac{n \sin a_n}{n \cos a_n}.$$

Consequently, $\tan a_1 < \dfrac{\sin a_1 + \sin a_2 + ... + \sin a_n}{\cos a_1 + \cos a_2 + ... + \cos a_n} < \tan a_n.$

Problem 54. Simplify $S = \dfrac{1}{\sin 2\theta} + \dfrac{1}{\sin 2^2\theta} + \dfrac{1}{\sin 2^3\theta} + ... + \dfrac{1}{\sin 2^n\theta}.$

Solution. Simplify S.
Observe that
$$\frac{1}{\sin 2\alpha} = \frac{\cos^2\alpha + \sin^2\alpha}{\sin 2\alpha}$$
$$= \frac{2\cos^2\alpha - (\cos^2\alpha - \sin^2\alpha)}{\sin 2\alpha}$$
$$= \frac{2\cos^2\alpha}{2\sin\alpha\cos\alpha} - \frac{\cos 2\alpha}{\sin 2\alpha} = \cot\alpha - \cot 2\alpha.$$

Consequently,
$$\frac{1}{\sin 2\theta} = \cot\theta - \cot 2\theta$$
$$\frac{1}{\sin 2^2\theta} = \cot 2\theta - \cot 2^2\theta$$
$$\frac{1}{\sin 2^3\theta} = \cot 2^2\theta - \cot 2^3\theta$$
$$\vdots$$

and $\quad \dfrac{1}{\sin 2^n\theta} = \cot 2^{n-1}\theta - \cot 2^n\theta.$

Adding the equalities, we obtain $S = \cot\theta - \cot 2^n\theta$.
Therefore, $S = \cot\theta - \cot 2^n\theta$.

Problem 55. Find the product
$$P = \left(1 - \tan^2\frac{\theta}{2}\right)\left(1 - \tan^2\frac{\theta}{2^2}\right)\cdots\left(1 - \tan^2\frac{\theta}{2^n}\right).$$

Solution. Find the product P.
Using double angle formula, $\tan 2\alpha = \dfrac{2\tan\alpha}{1-\tan^2\alpha}$, we obtain
$$1 - \tan^2\alpha = \frac{2\tan\alpha}{\tan 2\alpha}.$$

Then
$$1 - \tan^2\frac{\theta}{2} = \frac{2\tan\dfrac{\theta}{2}}{\tan\theta}$$

$$1 - \tan^2\frac{\theta}{2^2} = \frac{2\tan\dfrac{\theta}{2^2}}{\tan\dfrac{\theta}{2}}$$

$$\vdots$$

$$\text{and} \quad 1 - \tan^2 \frac{\theta}{2^n} = \frac{2\tan \frac{\theta}{2^n}}{\tan \frac{\theta}{2^{n-1}}}.$$

Consequently,

$$P = \left(\frac{2\tan \frac{\theta}{2}}{\tan \theta}\right)\left(\frac{2\tan \frac{\theta}{2^2}}{\tan \frac{\theta}{2}}\right) \times \ldots \times \left(\frac{2\tan \frac{\theta}{2^n}}{\tan \frac{\theta}{2^{n-1}}}\right)$$

$$= \frac{2^n \tan \frac{\theta}{2^n}}{\tan \theta}.$$

Therefore, $P = \dfrac{2^n \tan \frac{\theta}{2^n}}{\tan \theta}$.

Problem 56. Simplify the following expressions:

1. $S_1 = 1 + \dfrac{\cos \theta}{\cos \theta} + \dfrac{\cos 2\theta}{\cos^2 \theta} + \ldots + \dfrac{\cos n\theta}{\cos^n \theta};$

2. $S_2 = \dfrac{\sin \theta}{\cos \theta} + \dfrac{\sin 2\theta}{\cos^2 \theta} + \ldots + \dfrac{\sin n\theta}{\cos^n \theta}.$

Solution. Simplify the following expressions:

1. $S_1 = 1 + \dfrac{\cos \theta}{\cos \theta} + \dfrac{\cos 2\theta}{\cos^2 \theta} + \ldots + \dfrac{\cos n\theta}{\cos^n \theta}$
   Observe that

$$\frac{\sin(k+1)\theta}{\cos^k \theta} = \frac{\sin k\theta \cos \theta + \sin \theta \cos k\theta}{\cos^k \theta}$$

$$= \frac{\sin k\theta \cos x}{\cos^k \theta} + \frac{\sin \theta \cos k\theta}{\cos^k \theta}$$

$$= \frac{\sin k\theta}{\cos^{k-1}\theta} + \frac{\sin \theta \cos k\theta}{\cos^k \theta}.$$

Then

$$\frac{\sin \theta \cos k\theta}{\cos^k \theta} = \frac{\sin(k+1)\theta}{\cos^k \theta} - \frac{\sin k\theta}{\cos^{k-1}\theta}$$

or

$$\frac{\cos k\theta}{\cos^k \theta} = \frac{1}{\sin \theta}\left[\frac{\sin(k+1)\theta}{\cos^k \theta} - \frac{\sin k\theta}{\cos^{k-1}\theta}\right].$$

We obtain

$$\frac{\cos\theta}{\cos\theta} = \frac{1}{\sin\theta}\left(\frac{\sin 2\theta}{\cos\theta} - \frac{\sin\theta}{\cos^0\theta}\right)$$

$$\frac{\cos 2\theta}{\cos^2\theta} = \frac{1}{\sin\theta}\left(\frac{\sin 3\theta}{\cos^2\theta} - \frac{\sin 2\theta}{\cos\theta}\right)$$

$$\frac{\cos 3\theta}{\cos^3\theta} = \frac{1}{\sin\theta}\left(\frac{\sin 4\theta}{\cos^3\theta} - \frac{\sin 3\theta}{\cos^2\theta}\right)$$

$$\vdots$$

$$\frac{\cos n\theta}{\cos^n\theta} = \frac{1}{\sin\theta}\left[\frac{\sin(n+1)\theta}{\cos^n\theta} - \frac{\sin n\theta}{\cos^{n-1}\theta}\right].$$

Adding the equalities, we obtain

$$S_1 = 1 + \frac{1}{\sin\theta}\left[\frac{\sin(n+1)\theta}{\cos^n\theta} - \frac{\sin\theta}{\cos^0\theta}\right]$$

$$= 1 + \frac{\sin(n+1)\theta}{\sin\theta\cos^n\theta} - 1 = \frac{\sin(n+1)\theta}{\sin\theta\cos^n\theta}.$$

Therefore, $S_1 = 1 + \dfrac{\sin(n+1)\theta}{\sin\theta\cos^n\theta} - 1 = \dfrac{\sin(n+1)\theta}{\sin\theta\cos^n\theta}.$

2. $S_2 = \dfrac{\sin\theta}{\cos\theta} + \dfrac{\sin 2\theta}{\cos^2\theta} + ... + \dfrac{\sin n\theta}{\cos^n\theta}.$

Observe that

$$\frac{\cos(k+1)\theta}{\cos^k\theta} = \frac{\cos k\theta\cos\theta - \sin k\theta\sin\theta}{\cos^k\theta}$$

$$= \frac{\cos k\theta\cos\theta}{\cos^k\theta} - \frac{\sin k\theta\sin\theta}{\cos^k\theta}.$$

Then

$$\frac{\sin k\theta\sin x}{\cos^k\theta} = \frac{\cos k\theta}{\cos^{k-1}\theta} - \frac{\cos(k+1)\theta}{\cos^k\theta}$$

or

$$\frac{\sin k\theta}{\cos^k\theta} = \frac{1}{\sin\theta}\left[\frac{\cos k\theta}{\cos^{k-1}\theta} - \frac{\cos(k+1)\theta}{\cos^k\theta}\right].$$

We obtain

$$\frac{\sin\theta}{\cos\theta} = \frac{1}{\sin\theta}\left(\frac{\cos\theta}{\cos^0\theta} - \frac{\cos 2\theta}{\cos\theta}\right)$$

182

$$\frac{\sin 2\theta}{\cos^2 \theta} = \frac{1}{\sin \theta} \left(\frac{\cos 2\theta}{\cos \theta} - \frac{\cos 3\theta}{\cos^2 \theta} \right)$$

$$\frac{\sin 3\theta}{\cos^3 \theta} = \frac{1}{\sin \theta} \left(\frac{\cos 3\theta}{\cos^2 \theta} - \frac{\cos 4\theta}{\cos^3 \theta} \right)$$

$$\vdots$$

$$\frac{\cos n\theta}{\cos^n \theta} = \frac{1}{\sin \theta} \left[\frac{\cos n\theta}{\cos^{n-1} \theta} - \frac{\cos (n+1)\theta}{\cos^n \theta} \right]$$

Adding the equalities, we obtain

$$S_2 = \frac{1}{\sin \theta} \left[\frac{\cos \theta}{\cos^0 \theta} - \frac{\cos (n+1)\theta}{\cos^n \theta} \right] = \frac{\cos \theta \cos^n \theta - \cos (n+1)\theta}{\sin \theta \cos^n \theta}.$$

Problem 57. Let r and R be the radii of the inscribed circle and the circumscribe circle of triangle ABC respectively. Prove that

$$\frac{1}{\sin \frac{A}{2}} + \frac{1}{\sin \frac{B}{2}} + \frac{1}{\sin \frac{C}{2}} \geq 4\sqrt{\frac{R}{r}}.$$

Solution. Prove that $\dfrac{1}{\sin \frac{A}{2}} + \dfrac{1}{\sin \frac{B}{2}} + \dfrac{1}{\sin \frac{C}{2}} \geq 4\sqrt{\dfrac{R}{r}}.$

Using law of cosine, $\cos A = \dfrac{b^2 + c^2 - a^2}{2bc}.$
Then

$$\sin^2 \frac{A}{2} = \frac{1 - \cos A}{2}$$

$$= \frac{1 - \dfrac{b^2 + c^2 - a^2}{2bc}}{2}$$

$$= \frac{2bc - b^2 - c^2 + a^2}{4bc}$$

$$= \frac{a^2 - (b^2 - 2bc + c^2)}{4bc}$$

$$= \frac{a^2 - (b - c)^2}{4bc}$$

$$= \frac{(a + b - c)(a - b + c)}{4bc}$$

$$= \frac{(p - c)(p - b)}{bc}.$$

183

Additionally, $0 < A < \pi$, then $\sin\dfrac{A}{2} > 0$.

We obtain $\sin\dfrac{A}{2} = \sqrt{\dfrac{(p-b)(p-c)}{bc}}$.

Similarly, $\sin\dfrac{B}{2} = \sqrt{\dfrac{(p-c)(p-a)}{ca}}$ and $\sin\dfrac{C}{2} = \sqrt{\dfrac{(p-a)(p-b)}{ab}}$.

It follows that

$$\sin\dfrac{A}{2}\sin\dfrac{B}{2}\sin\dfrac{C}{2}$$
$$= \sqrt{\dfrac{(p-b)(p-c)}{bc}} \times \sqrt{\dfrac{(p-c)(p-a)}{ca}} \times \sqrt{\dfrac{(p-a)(p-b)}{ab}}$$
$$= \dfrac{(p-a)(p-b)(p-c)}{abc}.$$

Moreover, from Heron's formula,

$$S = \sqrt{p(p-a)(p-b)(p-c)}$$

or

$$(p-a)(p-b)(p-c) = \dfrac{S^2}{p} = \dfrac{prS}{p} = rS.$$

Using $\dfrac{abc}{4R} = S$, it implies that $abc = 4RS$.

We obtain $\sin\dfrac{A}{2}\sin\dfrac{B}{2}\sin\dfrac{C}{2} = \dfrac{rS}{4RS} = \dfrac{r}{4R}$.

Then $\dfrac{R}{r} = \dfrac{1}{4\sin\frac{A}{2}\sin\frac{B}{2}\sin\frac{C}{2}}$.

The given inequality, $\dfrac{1}{\sin\frac{A}{2}} + \dfrac{1}{\sin\frac{B}{2}} + \dfrac{1}{\sin\frac{C}{2}} \geq 4\sqrt{\dfrac{R}{r}}$ is equivalent to

$$\sqrt{\dfrac{\sin\frac{B}{2}\sin\frac{C}{2}}{\sin\frac{A}{2}}} + \sqrt{\dfrac{\sin\frac{C}{2}\sin\frac{A}{2}}{\sin\frac{B}{2}}} + \sqrt{\dfrac{\sin\frac{A}{2}\sin\frac{B}{2}}{\sin\frac{C}{2}}} \geq 2. \qquad (1)$$

To show the claim, it is sufficient to show that (1) is true.
We have

$$\sin\dfrac{B}{2}\sin\dfrac{C}{2} = \sqrt{\dfrac{(p-c)(p-a)}{ca}} \times \sqrt{\dfrac{(p-a)(p-b)}{ab}}$$
$$= \left(\dfrac{p-a}{a}\right)\sqrt{\dfrac{(p-b)(p-c)}{bc}}$$

$$= \left(\frac{p-a}{a}\right)\sin\frac{A}{2}.$$

Then $\dfrac{\sin\frac{B}{2}\sin\frac{C}{2}}{\sin\frac{A}{2}} = \dfrac{p-a}{a}.$

Likewise, $\dfrac{\sin\frac{C}{2}\sin\frac{A}{2}}{\sin\frac{B}{2}} = \dfrac{p-b}{b}$ and $\dfrac{\sin\frac{A}{2}\sin\frac{B}{2}}{\sin\frac{C}{2}} = \dfrac{p-c}{c}.$

It implies that

$$\sqrt{\frac{\sin\frac{B}{2}\sin\frac{C}{2}}{\sin\frac{A}{2}}} + \sqrt{\frac{\sin\frac{C}{2}\sin\frac{A}{2}}{\sin\frac{B}{2}}} + \sqrt{\frac{\sin\frac{A}{2}\sin\frac{B}{2}}{\sin\frac{C}{2}}}$$
$$= \sqrt{\frac{p-a}{a}} + \sqrt{\frac{p-b}{b}} + \sqrt{\frac{p-c}{c}}.$$

Using AM-GM inequality, we obtain $\sqrt{(p-a)a} \le \dfrac{p-a+a}{2} = \dfrac{p}{2}.$

Multiply both sides of the inequality by $\dfrac{2}{p}\sqrt{\dfrac{p-a}{a}}$, it follows that

$$\left(\frac{p}{2}\right)\left(\frac{2}{p}\sqrt{\frac{p-a}{a}}\right) \ge \sqrt{(p-a)a}\left(\frac{2}{p}\sqrt{\frac{p-a}{a}}\right)$$

or

$$\sqrt{\frac{p-a}{a}} \ge 2\left(\frac{p-a}{p}\right).$$

Similarly, $\sqrt{\dfrac{p-b}{b}} \ge 2\left(\dfrac{p-b}{p}\right)$ and $\sqrt{\dfrac{p-c}{c}} \ge 2\left(\dfrac{p-c}{p}\right).$

Consequently,

$$\sqrt{\frac{\sin\frac{B}{2}\sin\frac{C}{2}}{\sin\frac{A}{2}}} + \sqrt{\frac{\sin\frac{C}{2}\sin\frac{A}{2}}{\sin\frac{B}{2}}} + \sqrt{\frac{\sin\frac{A}{2}\sin\frac{B}{2}}{\sin\frac{C}{2}}}$$
$$\ge 2\left(\frac{p-a}{p}\right) + 2\left(\frac{p-b}{p}\right) + 2\left(\frac{p-c}{p}\right)$$
$$= 2\left(\frac{p-a+p-b+p-c}{p}\right) = 2\left(\frac{3p-2p}{p}\right) = 2.$$

Thus, the claim is proved.

www.ingramcontent.com/pod-product-compliance
Lightning Source LLC
Chambersburg PA
CBHW031625210526
45464CB00004B/1761